MATH

This Book Belongs To:

_____

_____

_____

_____

MULTIPLICATION TABLE

① 1

1 x 1 = 1
1 x 2 = 2
1 x 3 = 3
1 x 4 = 4
1 x 5 = 5
1 x 6 = 6
1 x 7 = 7
1 x 8 = 8
1 x 9 = 9
1 x 10 = 10

② 2

2 x 1 = 2
2 x 2 = 4
2 x 3 = 6
2 x 4 = 8
2 x 5 = 10
2 x 6 = 12
2 x 7 = 14
2 x 8 = 16
2 x 9 = 18
2 x 10 = 20

③ 3

3 x 1 = 3
3 x 2 = 6
3 x 3 = 9
3 x 4 = 12
3 x 5 = 15
3 x 6 = 18
3 x 7 = 21
3 x 8 = 24
3 x 9 = 27
3 x 10 = 30

④ 4

4 x 1 = 4
4 x 2 = 8
4 x 3 = 12
4 x 4 = 16
4 x 5 = 20
4 x 6 = 24
4 x 7 = 28
4 x 8 = 32
4 x 9 = 36
4 x 10 = 40

⑤ 5

5 x 1 = 5
5 x 2 = 10
5 x 3 = 15
5 x 4 = 20
5 x 5 = 25
5 x 6 = 30
5 x 7 = 35
5 x 8 = 40
5 x 9 = 45
5 x 10 = 50

MULTIPLICATION TABLE

6
6 × 1 = 6
6 × 2 = 12
6 × 3 = 18
6 × 4 = 24
6 × 5 = 30
6 × 6 = 36
6 × 7 = 42
6 × 8 = 48
6 × 9 = 54
6 × 10 = 60

7
7 × 1 = 7
7 × 2 = 14
7 × 3 = 21
7 × 4 = 28
7 × 5 = 35
7 × 6 = 42
7 × 7 = 49
7 × 8 = 56
7 × 9 = 63
7 × 10 = 70

8
8 × 1 = 8
8 × 2 = 16
8 × 3 = 24
8 × 4 = 32
8 × 5 = 40
8 × 6 = 48
8 × 7 = 56
8 × 8 = 64
8 × 9 = 72
8 × 10 = 80

9
9 × 1 = 9
9 × 2 = 18
9 × 3 = 27
9 × 4 = 36
9 × 5 = 45
9 × 6 = 54
9 × 7 = 63
9 × 8 = 72
9 × 9 = 81
9 × 10 = 90

10
10 × 1 = 10
10 × 2 = 20
10 × 3 = 30
10 × 4 = 40
10 × 5 = 50
10 × 6 = 60
10 × 7 = 70
10 × 8 = 80
10 × 9 = 90
10 × 10 = 100

Addition Tables

ones	twos	threes	fours	fives	sixes
1+1=2	2+1=3	3+1=4	4+1=5	5+1=6	6+1=7
1+2=3	2+2=4	3+2=5	4+2=6	5+2=7	6+2=8
1+3=4	2+3=5	3+3=6	4+3=7	5+3=8	6+3=9
1+4=5	2+4=6	3+4=7	4+4=8	5+4=9	6+4=10
1+5=6	2+5=7	3+5=8	4+5=9	5+5=10	6+5=11
1+6=7	2+6=8	3+6=9	4+6=10	5+6=11	6+6=12
1+7=8	2+7=9	3+7=10	4+7=11	5+7=12	6+7=13
1+8=9	2+8=10	3+8=11	4+8=12	5+8=13	6+8=14
1+9=10	2+9=11	3+9=12	4+9=13	5+9=14	6+9=15
1+10=11	2+10=12	3+10=13	4+10=14	5+10=15	6+10=16
1+11=12	2+11=13	3+11=14	4+11=15	5+11=16	6+11=17
1+12=13	2+12=14	3+12=15	4+12=16	5+12=17	6+12=18

Addition Tables

sevens	eights	nines	tens	elevens	twelves
7+1=8	8+1=9	9+1=10	10+1=11	11+1=12	12+1=13
7+2=9	8+2=10	9+2=11	10+2=12	11+2=13	12+2=14
7+3=10	8+3=11	9+3=12	10+3=13	11+3=14	12+3=15
7+4=11	8+4=12	9+4=13	10+4=14	11+4=15	12+4=16
7+5=12	8+5=13	9+5=14	10+5=15	11+5=16	12+5=17
7+6=13	8+6=14	9+6=15	10+6=16	11+6=17	12+6=18
7+7=14	8+7=15	9+7=16	10+7=17	11+7=18	12+7=19
7+8=15	8+8=16	9+8=17	10+8=18	11+8=19	12+8=20
7+9=16	8+9=17	9+9=18	10+9=19	11+9=20	12+9=21
7+10=17	8+10=18	9+10=19	10+10=20	11+10=21	12+10=22
7+11=18	8+11=19	9+11=20	10+11=21	11+11=22	12+11=23
7+12=19	8+12=20	9+12=21	10+12=22	11+12=23	12+12=24

Subtraction Tables

ones	twos	threes	fours	fives	sixes
1-1=0	2-2=0	3-3=0	4-4=0	5-5=0	6-6=0
2-1=1	3-2=1	4-3=1	5-4=1	6-5=1	7-6=1
3-1=2	4-2=2	5-3=2	6-4=2	7-5=2	8-6=2
4-1=3	5-2=3	6-3=3	7-4=3	8-5=3	9-6=3
5-1=4	6-2=4	7-3=4	8-4=4	9-5=4	10-6=4
6-1=5	7-2=5	8-3=5	9-4=5	10-5=5	11-6=5
7-1=6	8-2=6	9-3=6	10-4=6	11-5=6	12-6=6
8-1=7	9-2=7	10-3=7	11-4=7	12-5=7	13-6=7
9-1=8	10-2=8	11-3=8	12-4=8	13-5=8	14-6=8
10-1=9	11-2=9	12-3=9	13-4=9	14-5=9	15-6=9
11-1=10	12-2=10	13-3=10	14-4=10	15-5=10	16-6=10
12-1=11	13-2=11	14-3=11	15-4=11	16-5=11	17-6=11

Subtraction Tables

sevens	eights	nines	tens	elevens	twelves
7−7=0	8−8=0	9−9=0	10−10=0	11−11=0	12−12=0
8−7=1	9−8=1	10−9=1	11−10=1	12−11=1	13−12=1
9−7=2	10−8=2	11−9=2	12−10=2	13−11=2	14−12=2
10−7=3	11−8=3	12−9=3	13−10=3	14−11=3	15−12=3
11−7=4	12−8=4	13−9=4	14−10=4	15−11=4	16−12=4
12−7=5	13−8=5	14−9=5	15−10=5	16−11=5	17−12=5
13−7=6	14−8=6	15−9=6	16−10=6	17−11=6	18−12=6
14−7=7	15−8=7	16−9=7	17−10=7	18−11=7	19−12=7
15−7=8	16−8=8	17−9=8	18−10=8	19−11=8	20−12=8
16−7=9	17−8=9	18−9=9	19−10=9	20−11=9	21−12=9
17−7=10	18−8=10	19−9=10	20−10=10	21−11=10	22−12=10
18−7=11	19−8=11	20−9=11	21−10=11	22−11=11	23−12=11

DIVISION FACTS

÷ BY1

$1 \div 1 = 1$

$2 \div 1 = 2$

$3 \div 1 = 3$

$4 \div 1 = 4$

$5 \div 1 = 5$

$6 \div 1 = 6$

$7 \div 1 = 7$

$8 \div 1 = 8$

$9 \div 1 = 9$

$10 \div 1 = 10$

÷ BY2

$2 \div 2 = 1$

$4 \div 2 = 2$

$6 \div 2 = 3$

$8 \div 2 = 4$

$10 \div 2 = 5$

$12 \div 2 = 6$

$14 \div 2 = 7$

$16 \div 2 = 8$

$18 \div 2 = 9$

$20 \div 2 = 10$

÷ BY3

$3 \div 3 = 1$

$6 \div 3 = 2$

$9 \div 3 = 3$

$12 \div 3 = 4$

$15 \div 3 = 5$

$18 \div 3 = 6$

$21 \div 3 = 7$

$24 \div 3 = 8$

$27 \div 3 = 9$

$30 \div 3 = 10$

÷ BY4

$4 \div 4 = 1$

$8 \div 4 = 2$

$12 \div 4 = 3$

$16 \div 4 = 4$

$20 \div 4 = 5$

$24 \div 4 = 6$

$28 \div 4 = 7$

$32 \div 4 = 8$

$36 \div 4 = 9$

$40 \div 4 = 10$

÷ BY5

$5 \div 5 = 1$

$10 \div 5 = 2$

$15 \div 5 = 3$

$20 \div 5 = 4$

$25 \div 5 = 5$

$30 \div 5 = 6$

$35 \div 5 = 7$

$40 \div 5 = 8$

$45 \div 5 = 9$

$50 \div 5 = 10$

÷ BY6

$6 \div 6 = 1$

$12 \div 6 = 2$

$18 \div 6 = 3$

$24 \div 6 = 4$

$30 \div 6 = 5$

$36 \div 6 = 6$

$42 \div 6 = 7$

$48 \div 6 = 8$

$54 \div 6 = 9$

$60 \div 6 = 10$

DIVISION FACTS

÷ BY 7

$7 \div 7 = 1$

$14 \div 7 = 2$

$21 \div 7 = 3$

$28 \div 7 = 4$

$35 \div 7 = 5$

$42 \div 7 = 6$

$49 \div 7 = 7$

$56 \div 7 = 8$

$63 \div 7 = 9$

$70 \div 7 = 10$

÷ BY 8

$8 \div 8 = 1$

$16 \div 8 = 2$

$24 \div 8 = 3$

$32 \div 8 = 4$

$40 \div 8 = 5$

$48 \div 8 = 6$

$56 \div 8 = 7$

$64 \div 8 = 8$

$72 \div 8 = 9$

$80 \div 8 = 10$

÷ BY 9

$9 \div 9 = 1$

$18 \div 9 = 2$

$27 \div 9 = 3$

$36 \div 9 = 4$

$45 \div 9 = 5$

$54 \div 9 = 6$

$63 \div 9 = 7$

$72 \div 9 = 8$

$81 \div 9 = 9$

$90 \div 9 = 10$

÷ BY 10

$10 \div 10 = 1$

$20 \div 10 = 2$

$30 \div 10 = 3$

$40 \div 10 = 4$

$50 \div 10 = 5$

$60 \div 10 = 6$

$70 \div 10 = 7$

$80 \div 10 = 8$

$90 \div 10 = 9$

$100 \div 10 = 10$

Test Your Color

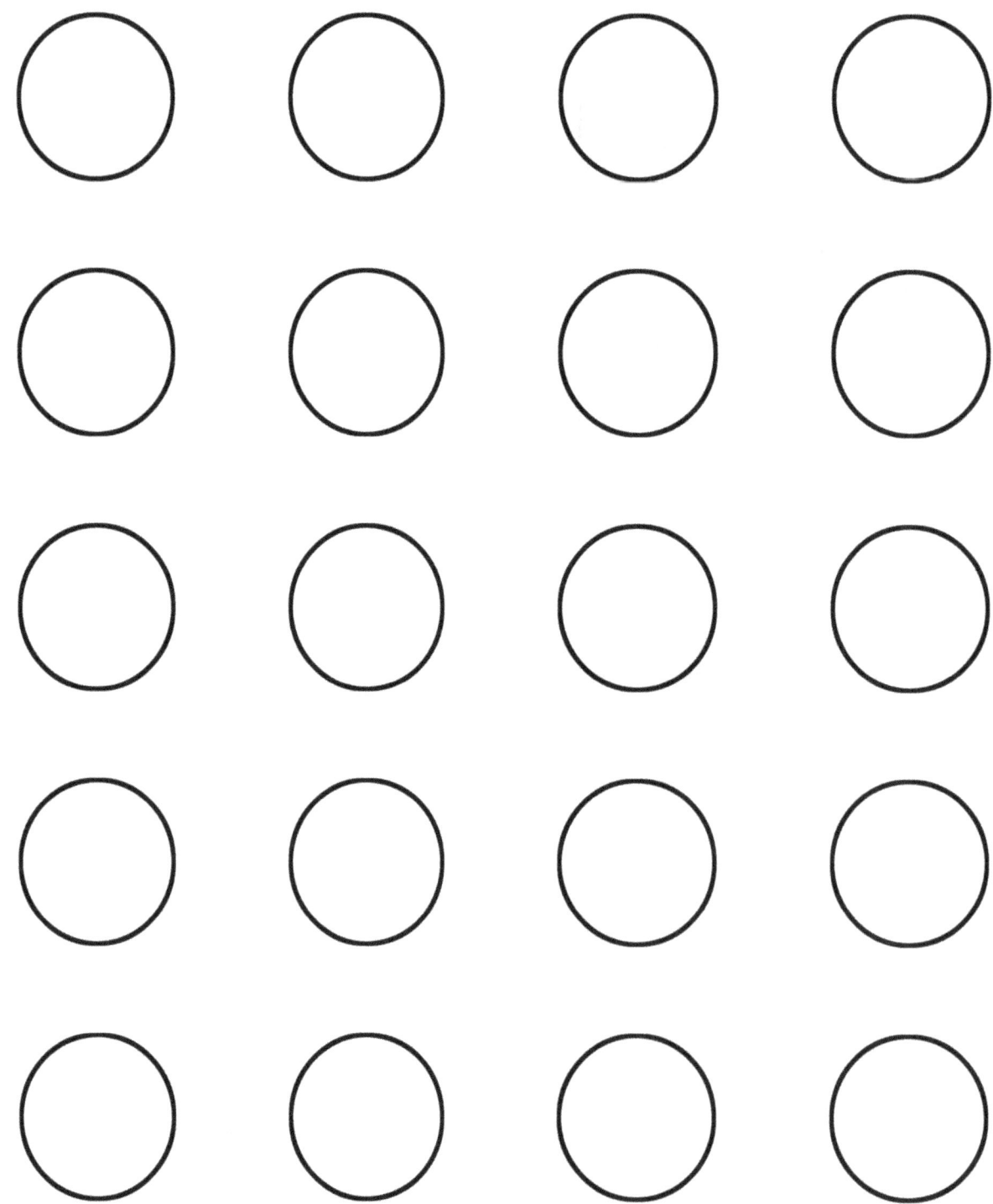

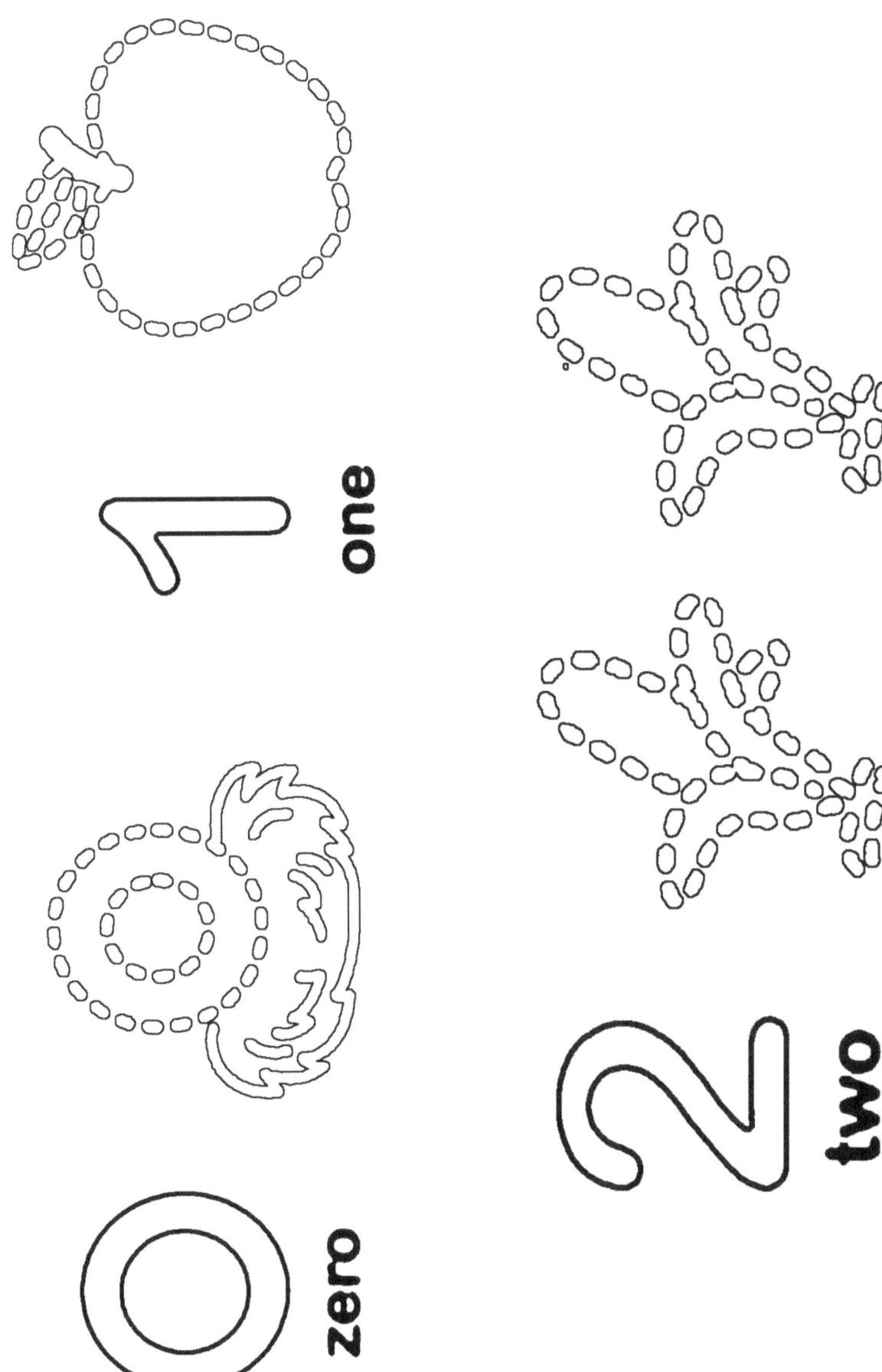

zero

one

two

Test Your Color

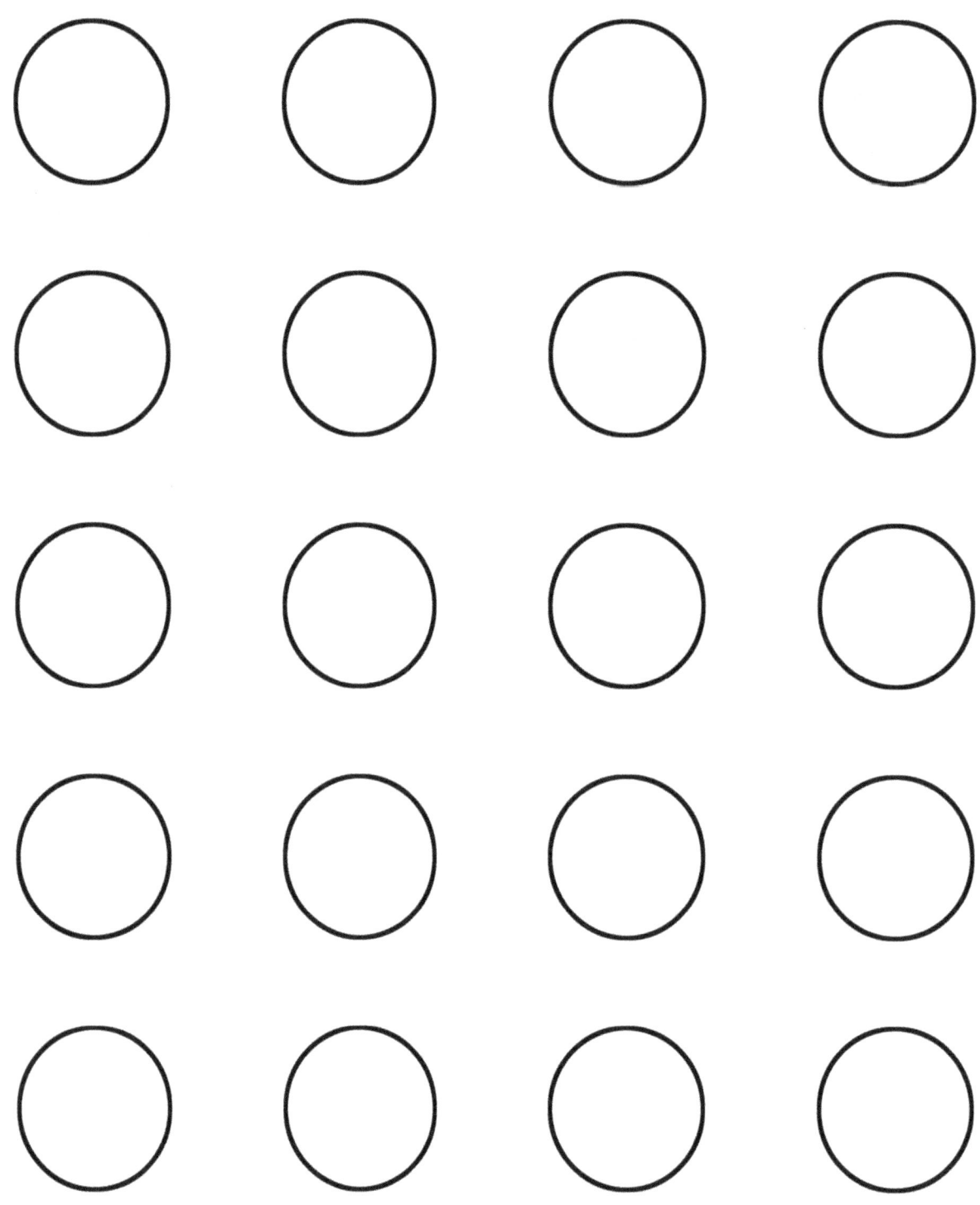

3
three

4
four

5
five

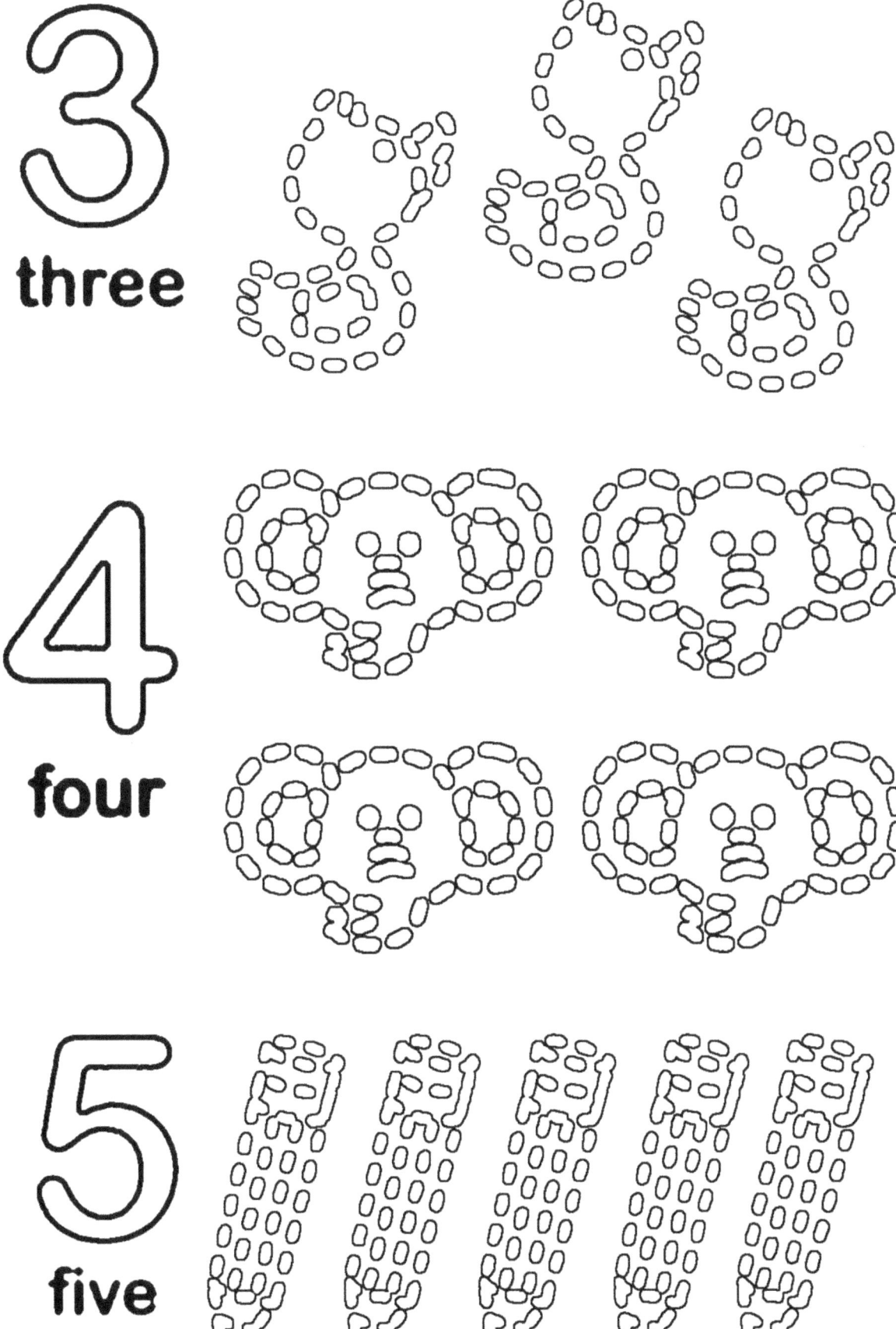

Test Your Color

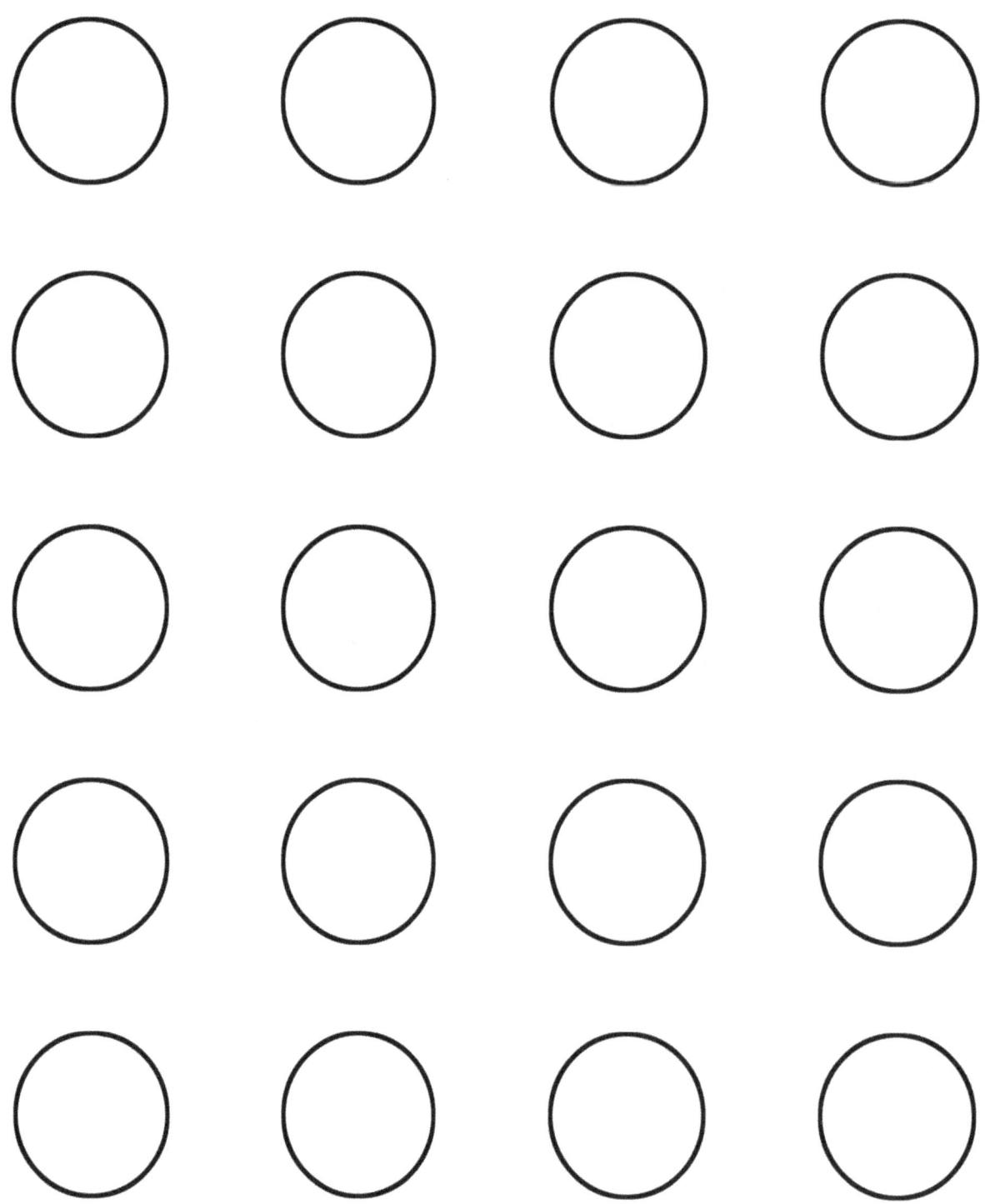

6
six

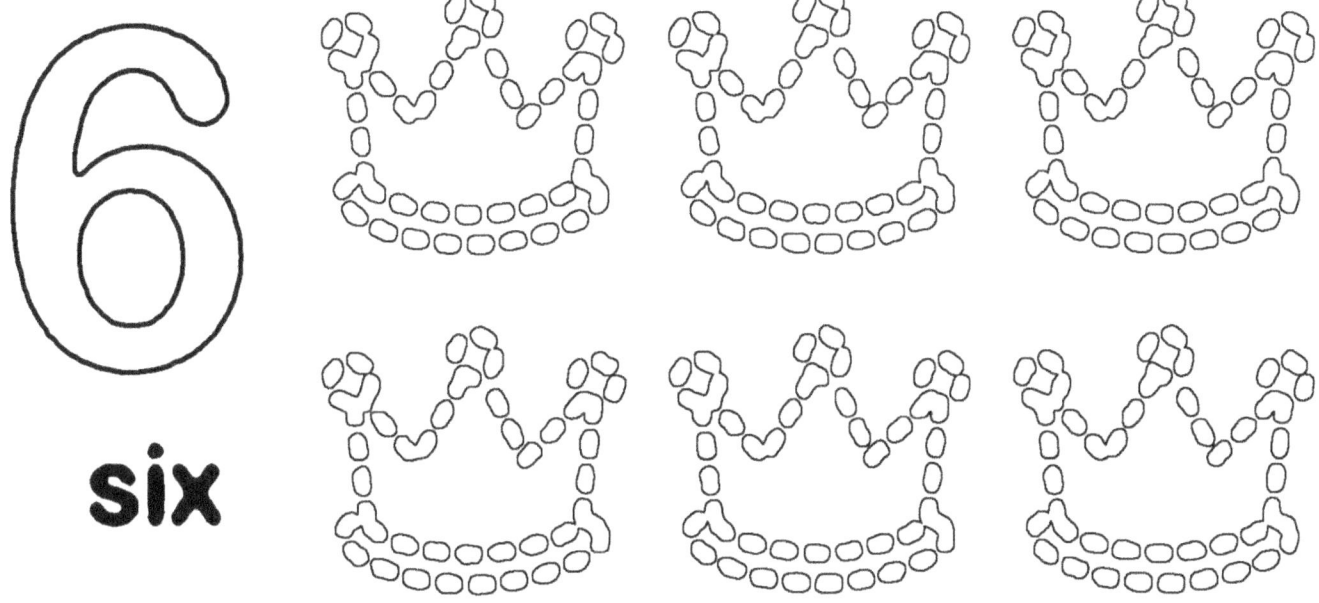

7
seven

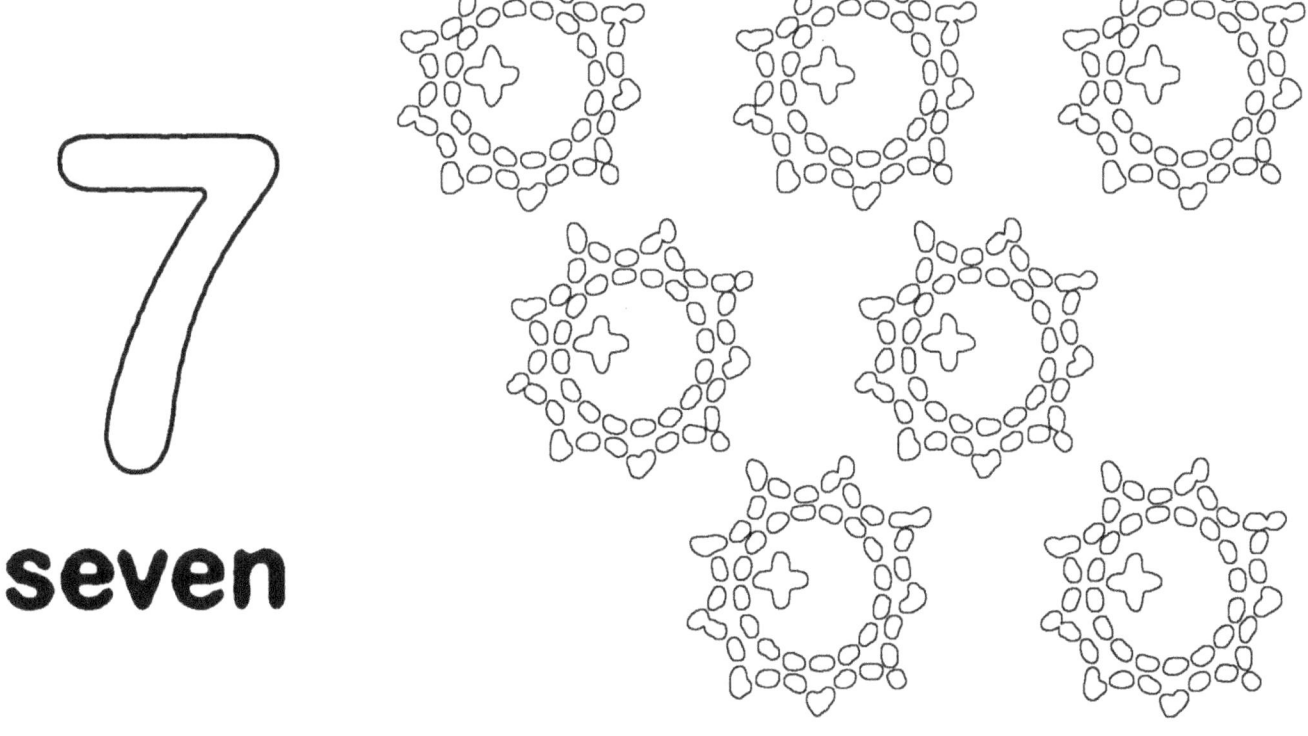

Test Your Color

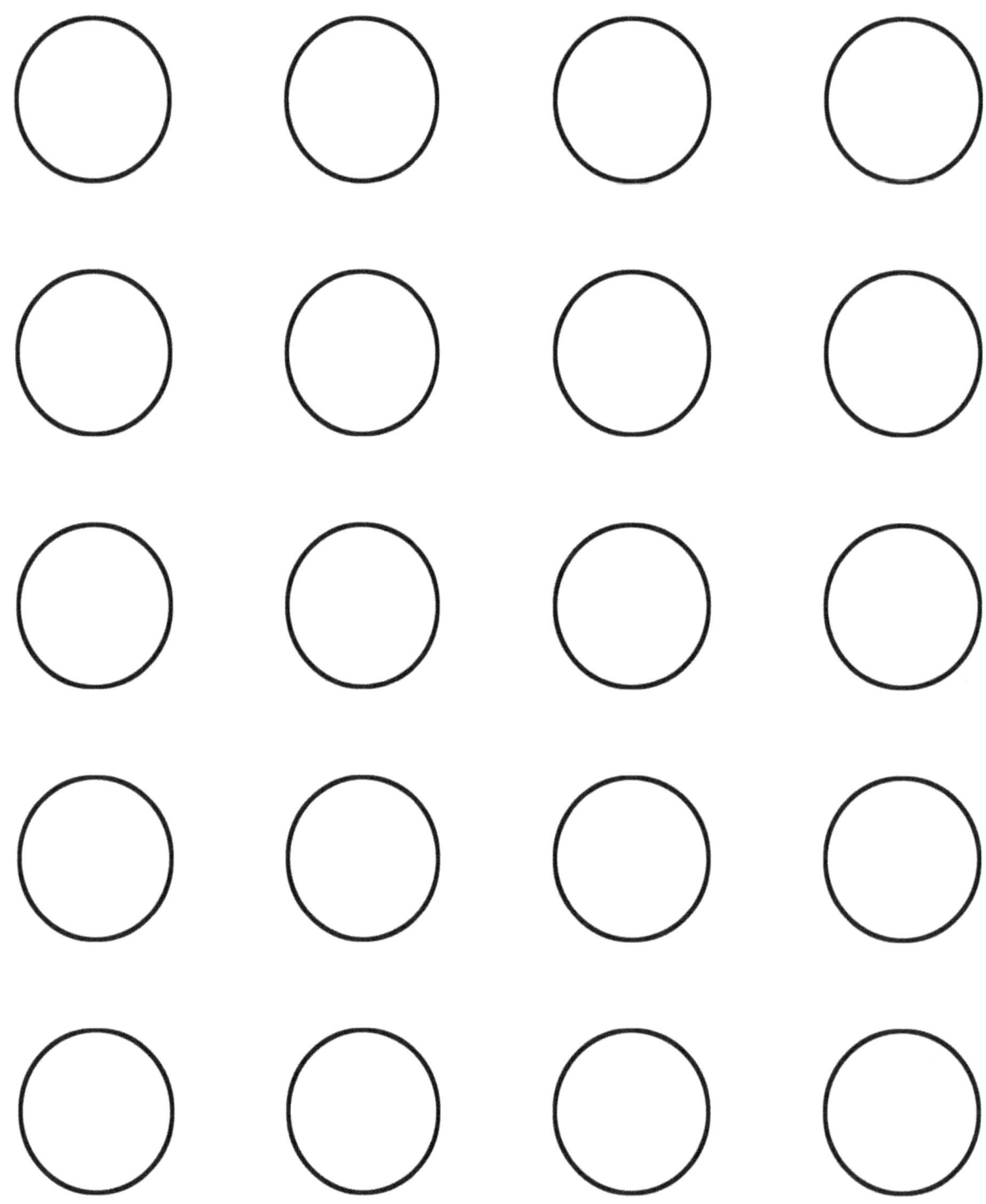

8
eight

9
nine

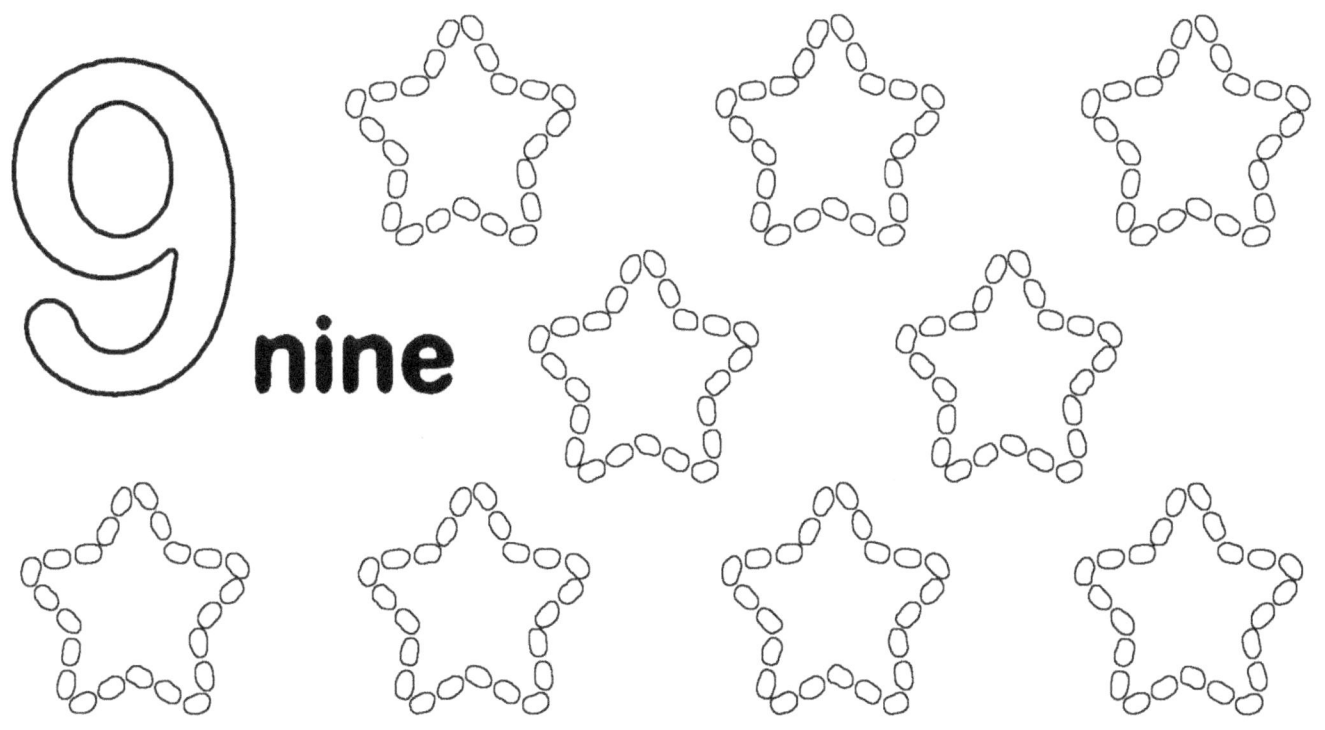

16 + 16 = ☐

25 + 29 = ☐

11 + 34 = ☐

78 + 12 = ☐

60 + 22 = ☐

54 − 24 = ☐

32 − 16 = ☐

89 − 55 = ☐

48 − 45 = ☐

39 − 12 = ☐

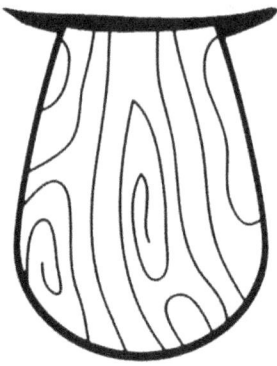
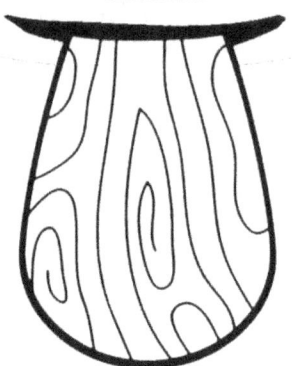

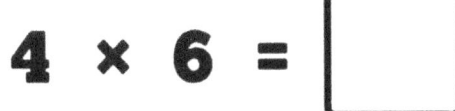

$4 \times 6 = \boxed{}$

$5 \times 3 = \boxed{}$

$7 \times 4 = \boxed{}$

$5 \times 8 = \boxed{}$

$9 \times 3 = \boxed{}$

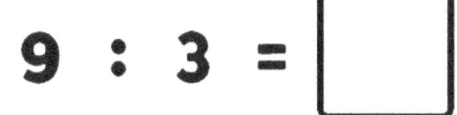

$9 : 3 = \boxed{}$

$48 : 6 = \boxed{}$

$32 : 8 = \boxed{}$

$28 : 4 = \boxed{}$

$54 : 9 = \boxed{}$

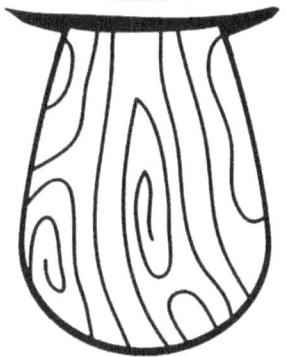

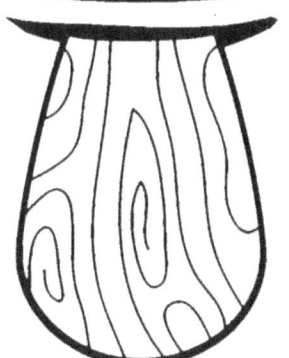

Math crossword

$1 + \square = 4$

5
$+$

7
$-$

$\square\ \square = \square\ 2$

2
$=$

$\square + 8 = \square$

$=$

$7 - \square = 4 + \square = 5$

$-$
5
$=$

$4 - \square = 0$

$+$
$2 + \square = \square$

$+$

$3 + \square = \square$ $9 - \square = 8$

$=$

$9 - \square = 6$

$+$

$=$

5

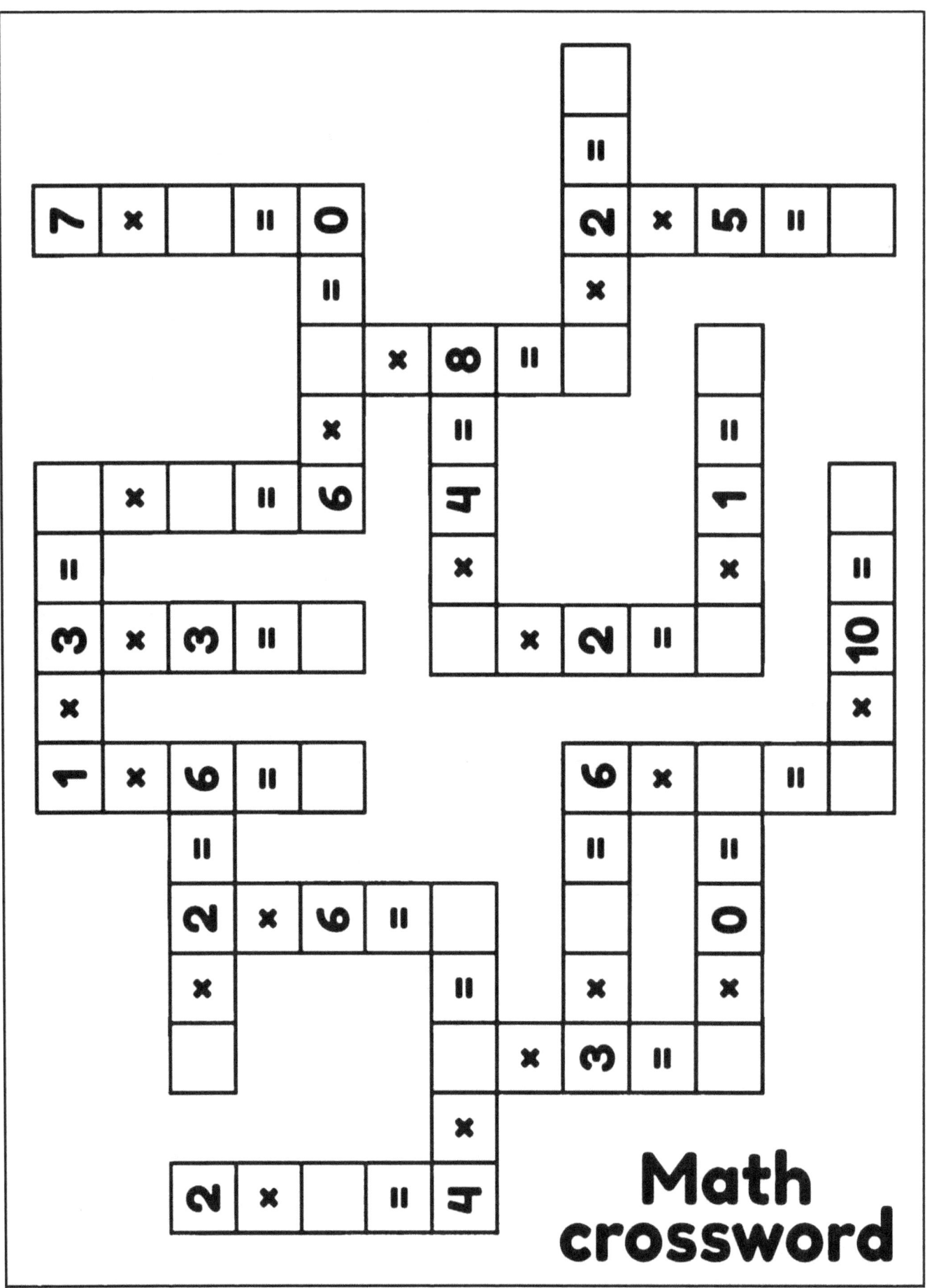

Math crossword

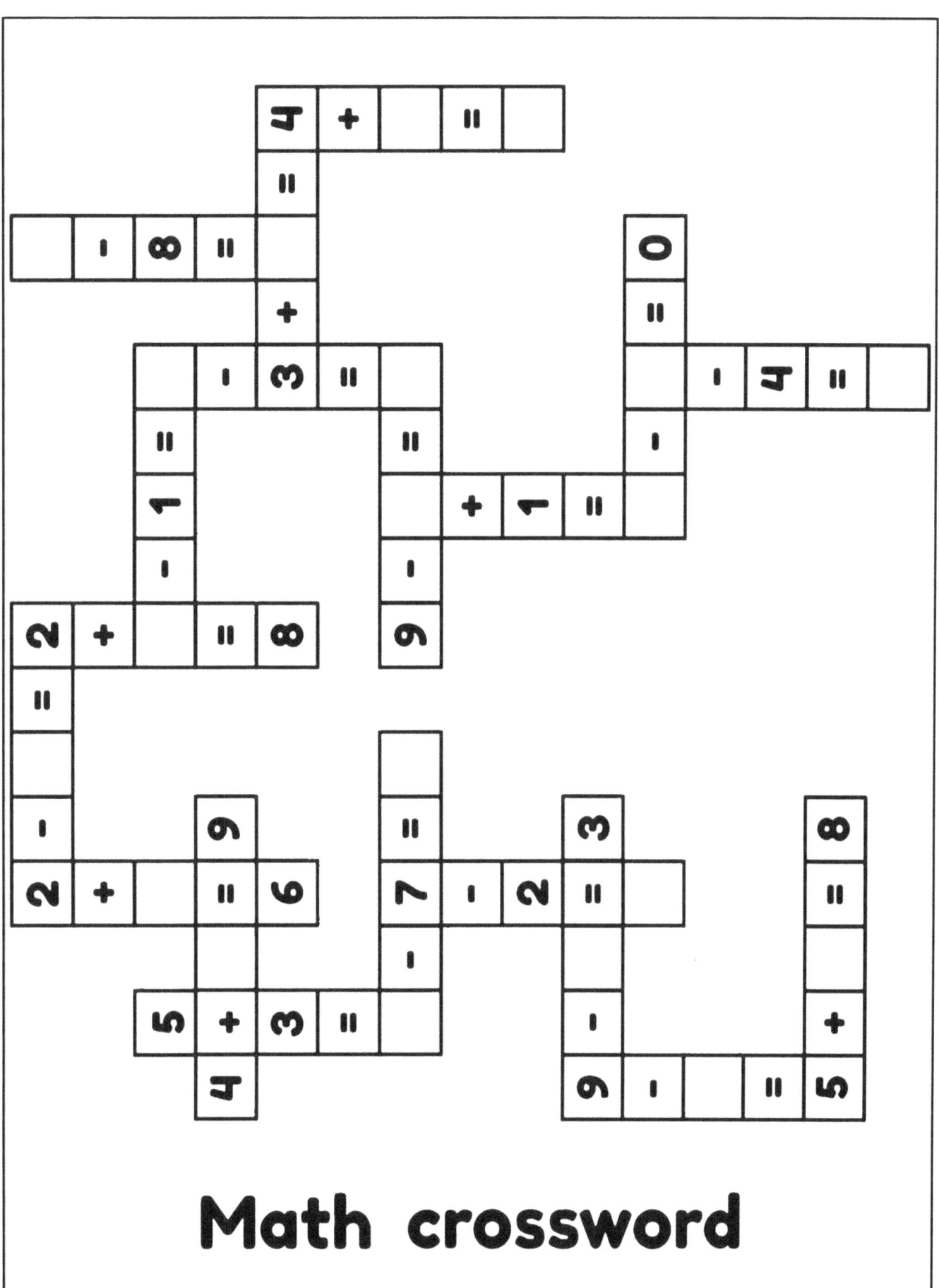

Math crossword

Math crossword

COMPLETE THE PATTERNS

Fill in the blank squares to complete each pattern.

10	11	10	11				
15	18			15			18
12	12				12		
	17	14			14	20	17
			13	16	13	16	
18	11	12				18	

COMPLETE THE PATTERNS

Fill in the blank squares to complete each pattern.

1	2	1	2				
3	4	5	3	4		3	
6	7	7	6		7		
		8	8	8			8
	9	0		0	9		
	2			5	2	5	2

FILL IN THE MISSING NUMBERS
TO COMPLETE THE SUMS

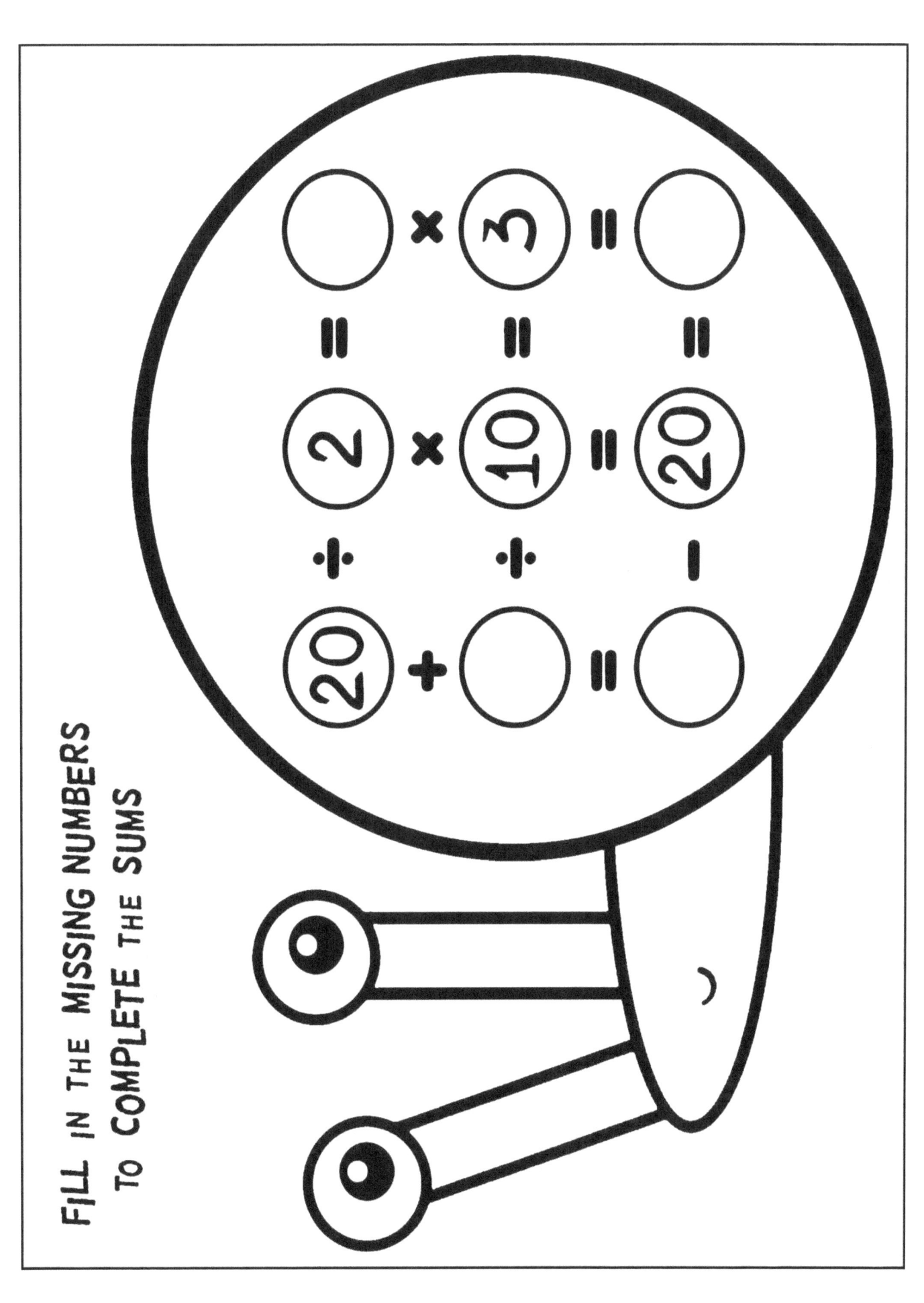

FILL IN THE MISSING NUMBERS TO COMPLETE THE SUMS

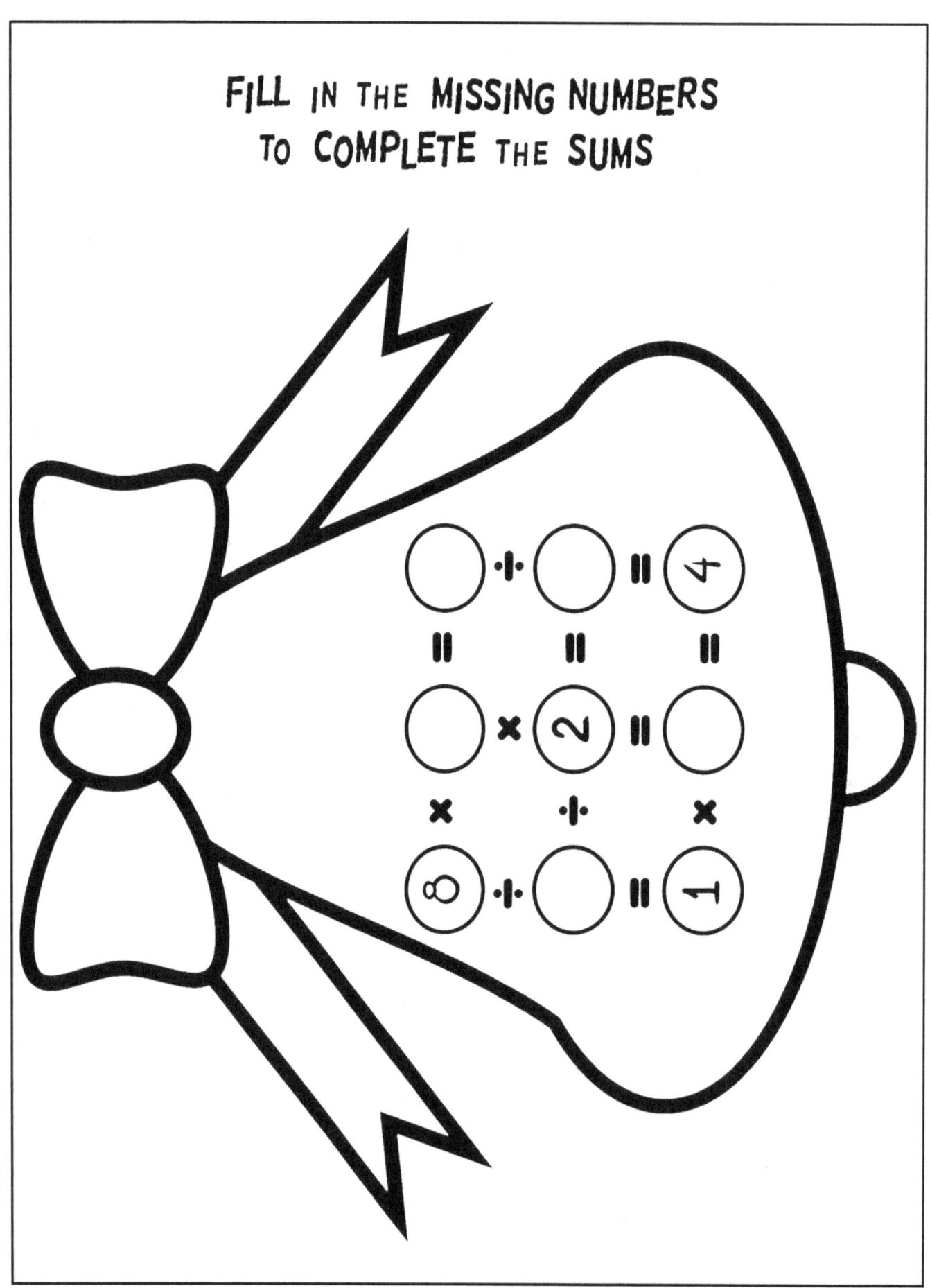

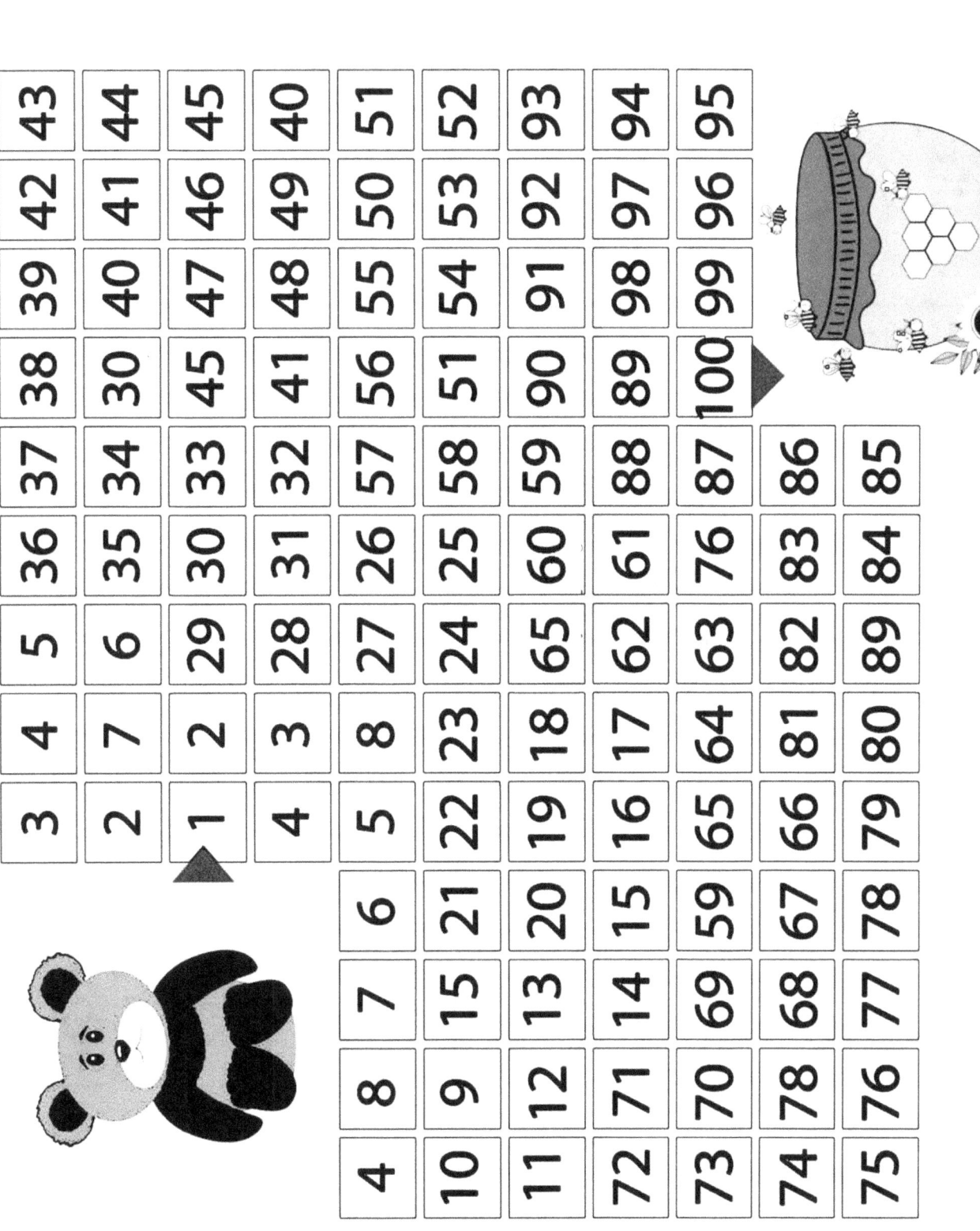

MATH MAZE +6

1	3	9	11	10
→	2	4	8	10

4	2	9	7	6	6	9
3	1	12	10	8	1	7
7	2	3	12	13	15	18
8	3	5	14	16	10	19
5	7	0	9	18	18	17
17	13	12	11	20		
14	13	10	12	→		

MATH MAZE FROM 1 TO 12

		1	2	4	8	9
	→	1	6	0	7	
2	3	4	2	3	5	1
8	9	11	6	4	2	3
7	5	7	6	5	10	5
0	9	8	4	2	1	0
5	7	9	10	1	11	2
4	3	8	11	9		
2	6	7	12	→		

2	3	4	5	5	6
1	8	9	6	11	12
		8	7	17	29
		9	22	23	24
25	26	10	28	29	30
31	12	11	34		
37	13	39	40		
43	14	15	46	47	20
49	50	16	17	18	19

Number Maze 1 - 10

10	9	7	6	4	
8	3	1	4	10	
10	4	2	1	8	9	10
3	5	7	3	7	2	1
2	1	2	10	6	5	2
	3	4	5	6	3
	4	7	9	7	1

MATH MAZE +1

7	9	8	10	7	6	5	4	2
10	8	6	5	4	3	2	1	⬆
12	9	7	9	5	6	5	3	2
13	10	8	9	10	11	12	13	9
20	17	18	14	11	12	13	14	15
26	21	23	26	20	14	19	13	15
23	30	27	26	25	20	19	14	16
⬆	29	28	30	24	23	22	21	20
30	28	24	25	28	24	26	23	27

10	12	11	13	10
16	14	12	10	13
17	19	11	13	16
18	14	20	15	14
19	15	28	11	12
25	23	20	19	16

1	10	11	8	9	2	1			
7	9	6	23	22	21	22▶			
7	6	5	7	8	20	6			
6	5	4	4	1	19	18	8	4	17
5	4	3	4	5	6	17	16	15	16
44	5	2	72	8	7	8	9	14	4
3	2	1▲	2	3	8	23	22	13	14
			11	10	9	10	11	12	5
			12	2	10	14	7	13	9
			13	39	11	12	13	14	15

			8	9	2	1	54	2	6
			1	2	3	12	6	8	1
			7	8	4	5	6	2	14
5	56	7	4	1	3	14	7	4	8
9	25	16	2	11	10	9	8	7	14
44	5	6	72	12	12	2	15	8	4
9	16	15	14	13	5	23	3	6	54
8	17	8	9	32	4	3			
38	18	19	1	23	24	25			
1	32	20	21	22	1	2			

Math Maze 12

0+12	6+6	7+5	8+4		10+2	11+1	12+0	↑	
1+11	10+1	1+7	4+5		9+3	3+11	7+3	8+8	6+4	6+5
2+10	3+9	4+8	5+7		6+7	8+6	9+3	8+4	7+5	8+5
7+7	8+1	5+5	7+8		12+0	11+1	10+2	8+7	6+6	7+4
9+6	4+9	8+3	10+2		2+11	3+6	5+9	3+10	0+12	8+9
	↑	8+4	9+3		11+1	12+0	9+7	2+10	1+11	2+5
					10+4	5+7	4+8	3+9	7+6	2+9

Math Maze 14

🍉	⬆	11+3	6+6	10+4	13+1	5+9	4+9	7+5	12+1
7+8	9+4	10+4	3+12	12+2	3+9	3+11	2+12	9+7	5+8
12+4	8+9	12+2	4+9	6+8	6+7	11+4	4+10	1+13	8+3
5+5	13+1	4+10	8+7	11+3	9+5	9+6	6+4	11+3	8+6
11+6	2+12	5+7	3+6	2+11	7+7	7+3	8+8	4+5	7+7
10+3	6+8	3+11	5+9	1+13	8+6	6+5	8+4	12+2	10+4
(hedgehog)					⬇	13+1	9+5	9+3	

MATH MAZE +3

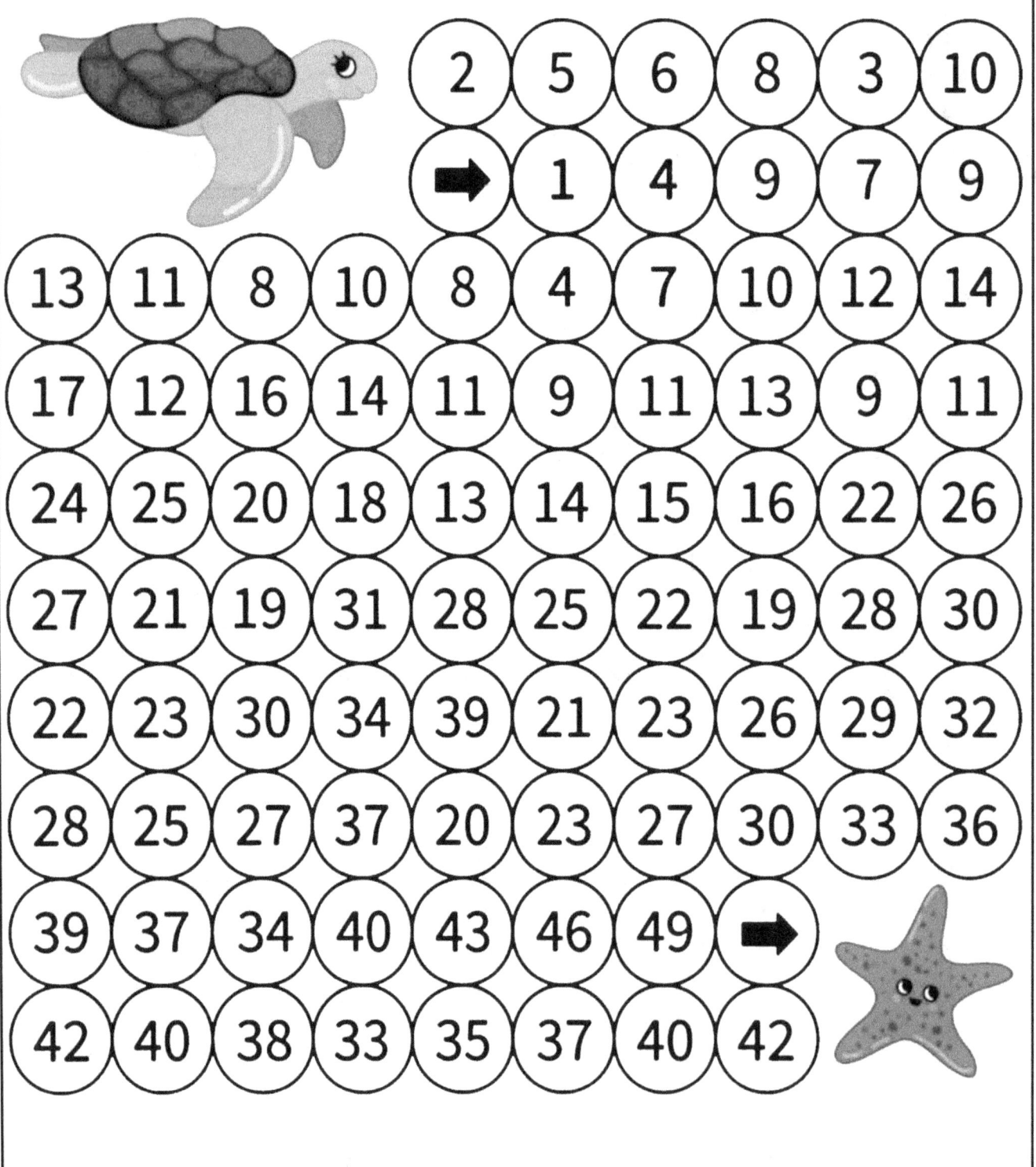

MATH MAZE +1

MATH MAZE FROM 1 TO 16

2	4	5	6	7
→ 1	0	5	8
2	4	3	2	3	8	9
6	5	4	5	6	9	8
10	6	7	13	10	15	19
11	3	8	9	12	11	20
12	10	9	16	18	13	19
7	11	12	15	16		
4	17	13	14	→		

				3	4	16	14	15	16	17	18	11
				2	5	6	13	12	25	24	19	20
				1	9	7	17	11	26	23	22	21
				46	45	8	9	10	27	20	31	32
53	52	51	48	47	44	43	42	13	28	29	30	33
54	45	50	49	76	77	49	41	40	39	38	35	34
55	56	44	74	75	78	79	80	81	76	37	36	38
51	57	58	73	72	71	76	83	82	86	98	99	33
61	60	59	68	69	70	89	84	95	96	97	100	86
62	56	54	67	88	87	86	85	94				
63	64	65	66	89	90	91	92	93				

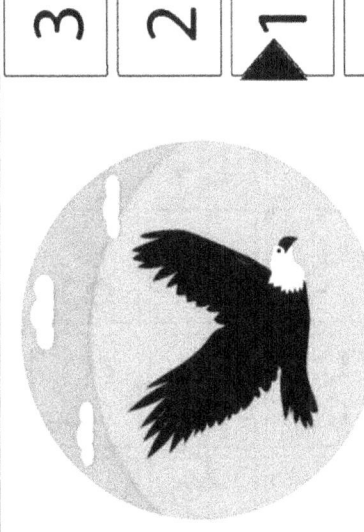

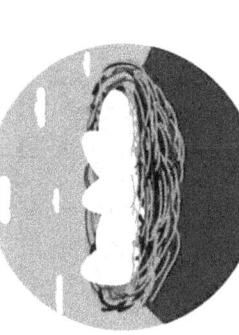

9	7	8	11	12	13	14	15	34	66	70	89	88
9	5	4	1	18	17	14	31	77	73	86

48	47	46	45	44	9	8	7	6	9
49	50	51	48	43	10	11	6	5	
58	57	52	53	42	18	12	3	4	
59	56	55	54	41	9	13	2	1	
60	61	36	39	40	19	14	6	18	19	20	21	22
65	62	63	38	37	36	15	16	17	28	27	29	23
61	68	64	65	39	35	34	33	14	29	26	25	24
99	98	97	66	67	68	66	32	31	30	77	78	79
100	94	96	95	94	69	70	71	77	75	76	85	80
				93	90	89	72	73	74	80	82	81
				92	91	88	87	86	85	84	83	87

17 12 10 16 13 12

21 20 19 18 17 10

10 18 12 15 16 12

16 12 19 14 16 17 14 11 19

10 15 11 17 11 12 13 15 17

4 7 8 9 10 9 14

9 5 7 5 7 2

4 8 6 2 3 9

6 9 5 1 6 7

8 2 4 3 2 1

+1

MATH MAZE +7

61	55	41	26	22	8	2 ↓	15	25	27
68	49	43	24	18	11	9	23	30	33
67	71	31	32	28	20	16	29	37	44
89	74	59	38	84	78	35	63	57	51
80	88	66	92	100	93	86	72	65	58
82	90	109	98	107	112	79	134	70	64
77	102	104	105	114	118	127	142 →		
83	91	96	111	121	128	135			

Math Maze 16

2+13	9+7	13+3	14+2	7+8		
15+1	11+5	12+3	10+6	6+4	↑	
7+9	4+9	5+7	1+15	13+1	2+14	
4+12	2+14	7+11	12+4	10+8	8+8	
3+8	5+11	7+7	7+9	4+7	5+11	
6+6	3+13	6+9	16+0	3+13	6+10	
↓	3+15	6+10	3+14	4+5	5+5	1+13
15+1	13+2	1+15	8+8	5+6	8+9	6+11
13+3	11+6	8+7	12+4	7+5	4+11	6+8
9+7	11+5	14+2	10+6	9+9	10+6	4+13

Math Maze 10

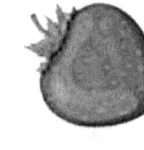
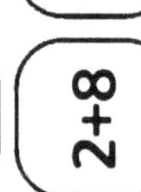

		0+10	9+1	8+2	7+3	2+6
		4+6	4+4	9+3	6+4	4+7
	2+8	3+7	5+6	5+5	0+10	2+8
	7+8	1+9	9+6	8+6	8+7	2+5
5+2	4+8	5+5	6+4	9+7	8+8	3+8
8+1	6+7	7+4	7+3	8+2	9+1	7+5
10+0	9+1	1+7	8+5	3+3	10+0	4+2
↑	8+2	6+6	2+8	4+6	5+5	7+6
	7+3	3+6	3+7	7+7	4+1	2+5
	6+4	5+5	1+9	4+5	3+5	6+2

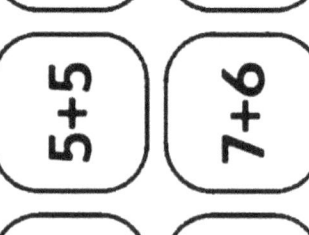

1	10	11	12	9	14	15			
7	9	6	23	22	13	22			
9	8	9	10	11	12	6			
6	7	6	4	1	13	14	8	30	17
5	4	5	6	5	14	15	16	29	28
2	3	4	5	6	7	16	17	18	27
1	2	5	20	19	18	17	22	13	26
			21	20	21	22	23	24	25
			22	2	22	14	7	13	26
			23	39	11	12	29	28	27

	1+2	2+3	4+4	1+8	3+4	🍎
	2+4	6+0	5+1	1+7	2+2	↑
	0+6	1+4	4+2	3+3	3+1	5+1
	1+5	4+3	3+2	1+5	6+2	6+0
	3+3	4+2	5+4	0+6	7+3	2+4
	4+1	5+1	3+7	2+4	5+3	0+6
↓	1+2	6+0	2+3	6+0	1+9	1+5
6+0	2+2	2+4	1+2	5+1	4+2	3+3
5+1	4+5	0+6	8+2	4+1	1+7	5+5
4+2	3+3	1+5	1+6	2+7	3+4	2+6

MATH MAZE +2

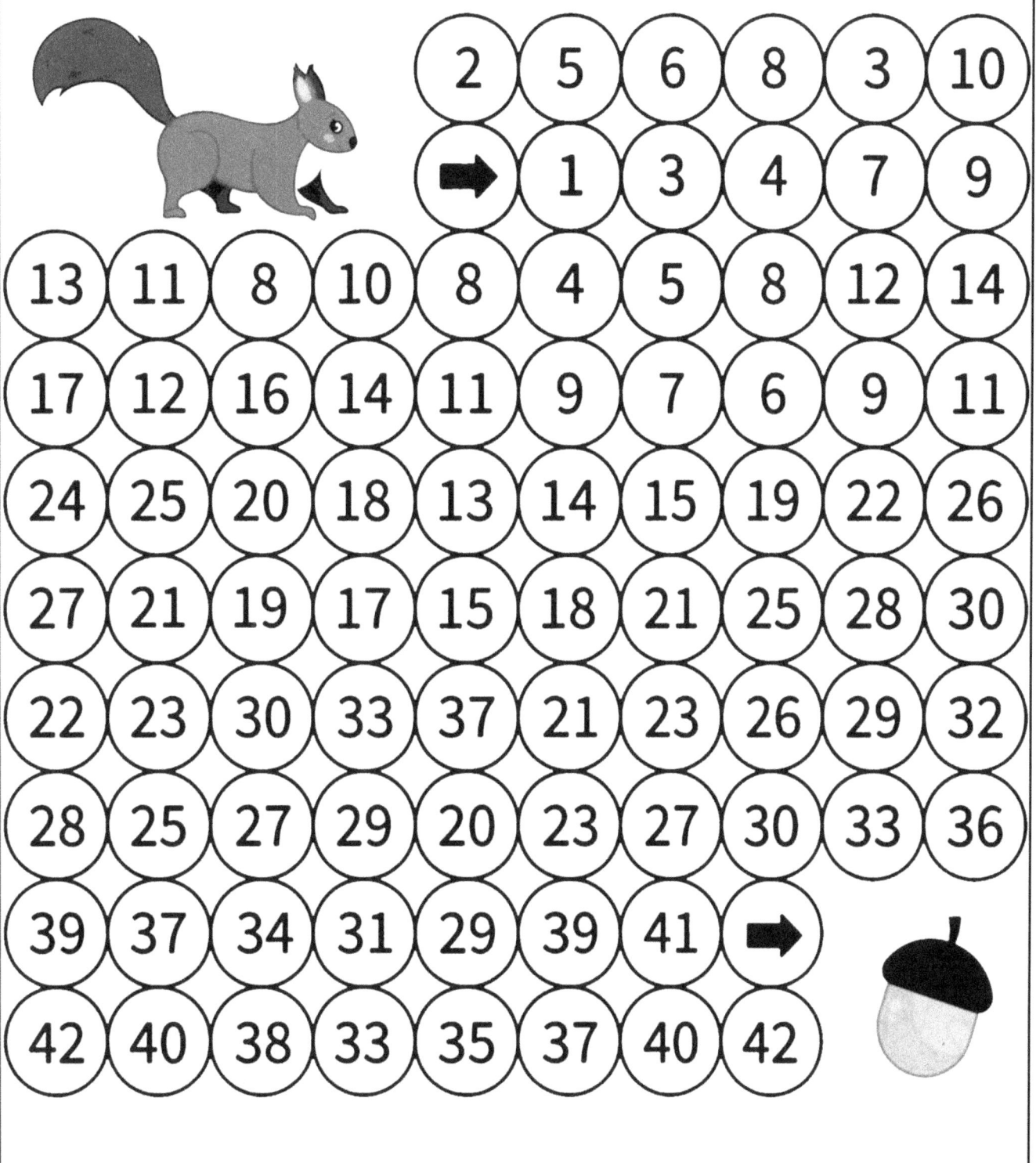

MATH MAZE +1

MATH MAZE +4

10	27	23	19	7	5	1	↓	🦔	↑ 🍓
12	25	21	17	13	9	3	8	2	
30	29	31	35	15	11	4	61	6	
39	33	37	41	45	49	53	57	65	69
14	26	32	28	34	82	59	55	43	73
72	52	46	40	38	42	83	89	47	77
50	54	58	51	79	60	94	93	85	81
56	66	62	78	80	88	96	97	98	84

Math Maze +2

Math Maze 11

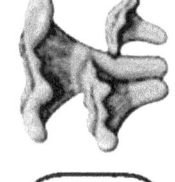

5+6	11+0	10+1	9+2	8+3		
4+7	5+5	6+2	2+5	7+4		
3+8	2+9	7+8	9+7	6+5	1+10	
8+2	0+11	1+10	3+5	9+1	6+7	
9+6	4+5	3+7	6+5	7+4	8+8	3+9
8+5	7+5	8+6	6+4	8+3	9+2	1+7
11+0	10+1	9+2	7+7	8+9	10+1	11+0
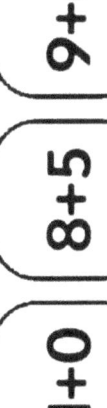	7+3	8+3	4+8	7+6	8+7	5+6
6+6	7+4	8+1	3+3	3+8	4+7	
8+1	6+5	1+10	0+11	2+9	3+6	

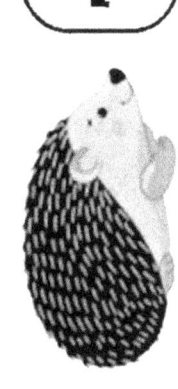

MATH MAZE +1

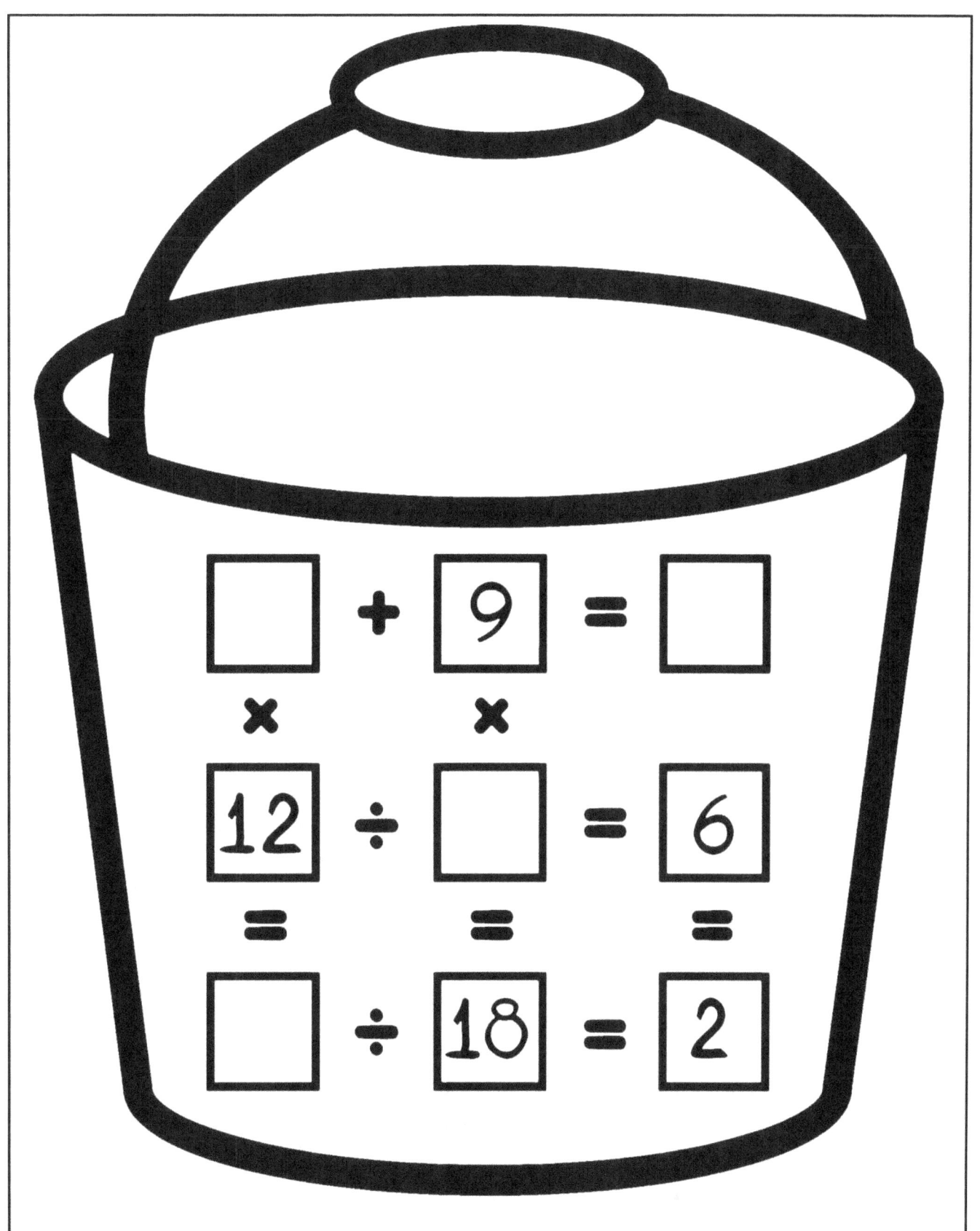

FILL IN THE MISSING NUMBERS
TO COMPLETE THE SUMS

Paste the missing numbers

1		3	4		6		8		10
11	12		14	15		17		19	20
	22	23	24		26		28	29	
31	32		34	35		37		39	40
41		43		45		47	48	49	
	52	53	54		56		58		60
61	62		64	65		67		69	70
	72	73		75	76		78	79	
81		83	84		86	87		89	90
	92	93		95	96		98		100

Paste the missing numbers

1	2		4		6	7	8		10
11	12	13		15	16	17		19	20
21	22		24	25			28	29	
31		33	34		36	37	38		40
	42	43		45	46		48	49	50
51	52	53	54		56		58		60
	62		64	65	66		68		70
71		73	74		76	77	78	79	
81	82		84		86	87		89	90
	92		94	95	96		98		100

Paste the missing numbers

1	2	3		5	6		8	9	10
11		13	14		16		18		20
21	22	23		25		27	28		30
	32		34		36			39	
	42			45	46		48		50
51			54	55		57	58		60
	62		64	65	66			69	
71		73			76	77	78		80
81		83	84		86	87		89	90
91	92		94		96		98		100

Paste the missing numbers

1	2		4	5	6	7		9	10
	12	13	14		16		18	19	
21		23	24	25		27	28		30
	32	33	34		36		38	39	
41			44		46		48		50
51	52	53		55	56	57	58	59	60
61		63	64		66	67		69	
	72	73	74	75		77	78	79	80
81	82	83		85		87	88		90
91		93	94	95	96	97		99	100

Paste the missing numbers

1		3	4	5		7	8	9	
	12	13	14		16	17	18		20
21	22	23		25	26	27		29	30
31	32		34	35	36		38	39	40
41		43	44	45		47	48	49	
	52	53	54		56	57	58		60
61	62	63		65	66	67		69	70
71	72		74	75	76		78	79	80
81		83	84	85		87	88	89	
	92	93	94		96	97	98		100

Paste the missing numbers

1		3		5		7		9	
11		13		15		17		19	
21		23		25		27		29	
31		33		35		37		39	
41		43		45		47		49	
51		53		55		57		59	
61		63		65		67		69	
71		73		75		77		79	
81		83		85		87		89	
91		93		95		97		99	

Paste the missing numbers

	2		4		6		8		10
11		13		15		17		19	
	22		24		26		28		30
31		33		35		37		39	
	42		44		46		48		50
51		53		55		57		59	
	62		64		66		68		70
71		73		75		77		79	
	82		84		86		88		90
91		93		95		97		99	

MATH MAZE

Help the two snowflakes get out of the maze. Make a path by drawing a line through the boxes that have sum of 50.

15 +29	15 +17	73 +12	22 +55	43 + 3	11 +44	11 +39 →
16 +61	16 +34	12 +55	44 + 6	23 + 8	4 +46	12 +38	78 +11
26 +24	26 +24	1 +49 50	←	54 +13	10 +40	28 +22
15 +35	16 +43	18 +15		2 +88	17 +28	17 +33
23 +27	42 + 8	38 +20	3 +54	19 + 8	25 +40	6 +44	16 +34
53 +15	20 +30	31 +19	36 +19	4 +46	24 +26	43 + 7	57 +16
19 +35	21 +13	5 +45	36 +14	4 +46	24 +21	57 +16	33 +23

MATH MAZE

Help the birds find the way to the birdhouses.
Make a path by drawing a line through the boxes
that have multiplication facts less than 50.

11 × 3 = 33	7 × 3	8 × 5	5 × 5	8 × 4	2 × 7
8 × 9	7 × 8	9 × 9	7 × 9	8 × 10	9 × 7	10 × 7	9 × 5
10 × 6	9 × 10	8 × 7			9 × 9	10 × 3	5 × 2
3 × 4	5 × 6	4 × 2			10 × 9	9 × 2	6 × 10
4 × 8	5 × 12	7 × 8	5 × 11	9 × 9	9 × 8	6 × 8	4 × 6
3 × 3	20 × 3	2 × 2	7 × 2	5 × 5	5 × 13	7 × 10	7 × 7
5 × 9	7 × 2	5 × 8	8 × 10	4 × 3	6 × 6	6 × 5	7 × 4

MATH MAZE

Help the bees find the way to the flowers. Make a path by drawing a line through the boxes that have sum of 25.

21 + 4 = 25	22 + 3	8 + 17	11 + 14	2 + 23	11 + 5
11 + 1	16 + 10	12 + 3	3 + 7	6 + 5	4 + 3	12 + 13	10 + 13
14 + 3	3 + 6	1 + 21		7 + 8	10 + 15	19 + 6
15 + 10	16 + 9	18 + 7		2 + 20	9 + 5	17 + 8
23 + 2	5 + 16	4 + 10	20 + 6	12 + 6	2 + 3	6 + 19	16 + 9
15 + 10	20 + 5	6 + 19	7 + 11	4 + 21	24 + 1	13 + 12	8 + 5
19 + 4	5 + 9	5 + 20	2 + 23	4 + 21	8 + 20	6 + 10	20 + 3

MATH MAZE

Help the rocket get out of the maze to the stars. Make a path
by drawing a line through the boxes that have sum of 30.

21 9	22 8	21 9	22 8	8 6	11 14		
11 19	16 10	12 3	3 27 __30__	6 5	4 3	12 18	10 13
14 16	3 27	1 21			7 8	10 20	19 11
15 7	16 14	18 7			2 20	9 5	17 13
23 10	5 25	4 10	20 6	12 6	2 3	6 24	16 14
15 10	20 10	6 24	7 23	4 21	24 1	13 17	8 5
19 4	5 9	5 20	2 28	4 26	8 22	6 24	20 3

MATH MAZE

Help the pencils find the way to the car and color it.
Make a path by drawing a line through the boxes
that have multiplication facts greater than 50.

10 × 4	7 × 2	11 × 3	7 × 3	8 × 5	5 × 5
8 × 9	7 × 8	9 × 9	7 × 7	8 × 4	9 × 2	10 × 8	9 × 5
10 × 6	9 × 4	8 × 7 = 56			9 × 5	10 × 7	5 × 19
6 × 12	5 × 6	4 × 2			10 × 3	9 × 2	6 × 10
4 × 23	5 × 12	7 × 3	5 × 7	9 × 2	9 × 4	6 × 10	11 × 5
3 × 3	10 × 6	2 × 2	7 × 2	5 × 13	5 × 13	7 × 10	7 × 7
5 × 9	7 × 9	5 × 16	8 × 10	4 × 19	6 × 6	6 × 5	7 × 4

Help the mouse get to the yummy slices of cheese.
Make a path by drawing a line through the boxes
that have difference of 2, 4 or 8.

16 − 8	10 − 8	14 − 10	16 − 12	19 − 18	17 − 3
19 − 15	8 − 1	12 − 6	9 − 1	16 − 14	11 − 7
14 − 6	15 − 7	14 − 13	10 − 4	8 − 3	4 − 2
15 − 1	19 − 11	13 − 7			12 − 4 = 8
18 − 11	5 − 3 = 2	15 − 9	9 − 8	15 − 5	14 − 7
		11 − 8	14 − 11	9 − 4	17 − 6

Help the butterflies get out of the maze.
Make a path by drawing a line through the boxes
that have difference of 3, 6 or 9.

11 − 1	17 − 13	19 − 17	11 − 3	7 − 1 6 →
14 − 9	16 − 8	12 − 5	10 − 9	9 − 3	13 − 4
15 − 12	← 14 − 8 6			12 − 7	17 − 8
10 − 1	14 − 3			18 − 17	9 − 6
8 − 2	10 − 4	6 − 2	18 − 8	16 − 7	8 − 5
12 − 11	15 − 9	4 − 1	19 − 10	18 − 9	7 − 6

MATH MAZE

Help the little snowman get out of the maze. Make a path by drawing a line through the boxes that have sum of 75.

32 +24	15 +60	73 + 2	22 +53	43 +32	11 +64	11 +64 
16 +59	16 +59	42 +25	38 +61	3 +28	19 +58	25 + 6	6 + 16
26 +49	19 +76	25 +37			36 +31	4 +71	24 + 17
15 +60	16 +43	18 +57			2 +73 75	17 + 58	17 + 58
23 +52	42 + 12	38 +20	3 +54	19 + 8	25 +40	6 + 42	16 + 59
53 +22	12 +38	31 +44	36 +39	4 +71	24 +51	43 +22	57 + 18
19 +56	21 +54	5 +70	36 +31	4 +79	24 +51	57 + 18	33 + 42

Help the butterflies get to the flowers.
Make a path by drawing a line through the boxes
that have sum of 8, 10 or 12.

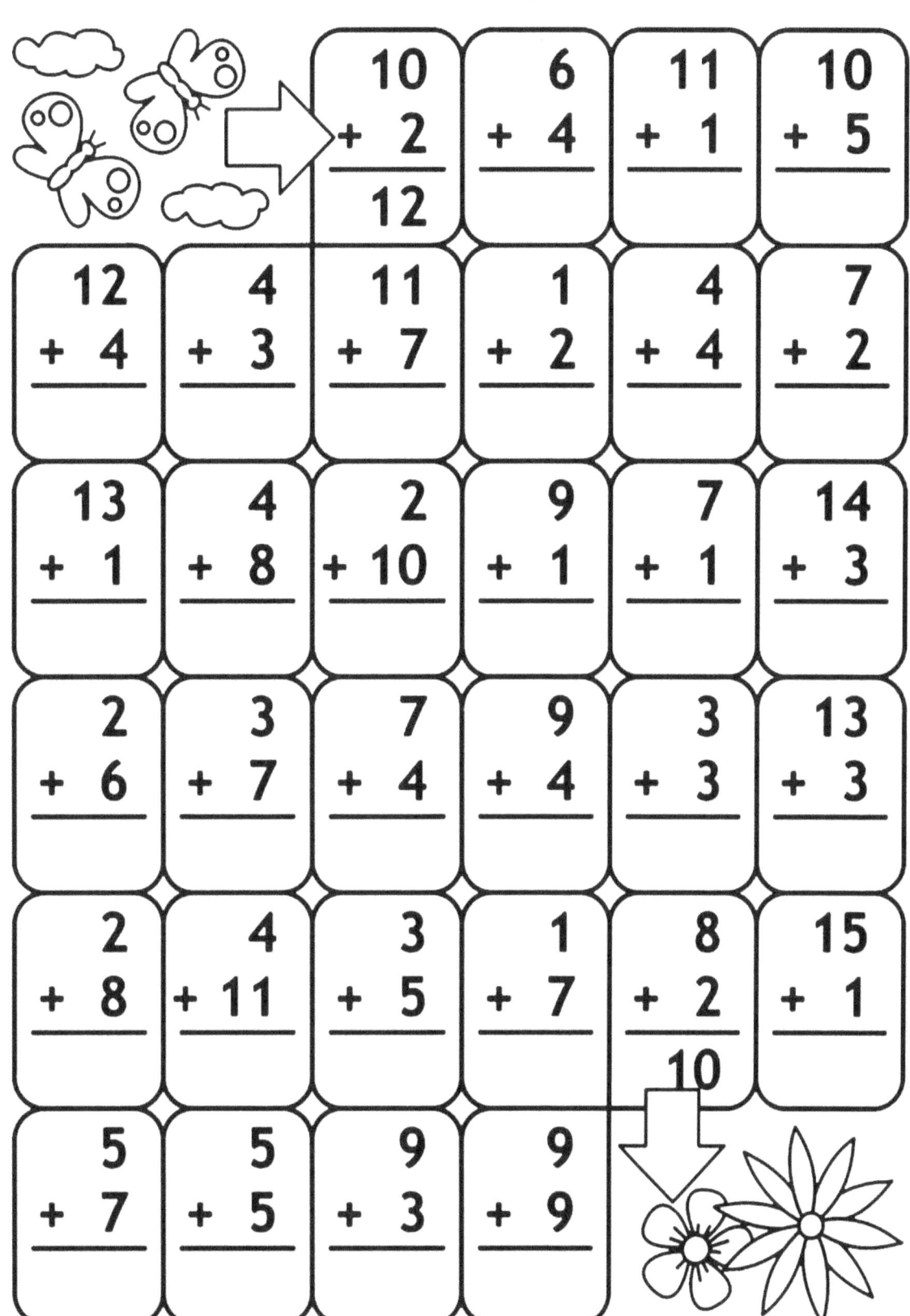

10 + 2 = 12	6 + 4	11 + 1	10 + 5
12 + 4	4 + 3	11 + 7	1 + 2	4 + 4	7 + 2
13 + 1	4 + 8	2 + 10	9 + 1	7 + 1	14 + 3
2 + 6	3 + 7	7 + 4	9 + 4	3 + 3	13 + 3
2 + 8	4 + 11	3 + 5	1 + 7	8 + 2 = 10	15 + 1
5 + 7	5 + 5	9 + 3	9 + 9

Help the snail get out of the maze.
Make a path by drawing a line through the boxes
that have sum of 7, 11 or 17.

10 + 1	14 + 3	3 + 8 —— 11	10 + 8	14 + 5	
1 + 6	12 + 2			9 + 3	16 + 1 —— 17
5 + 2	6 + 11	12 + 4	3 + 3	6 + 1	15 + 2
11 + 1	2 + 15	2 + 9	3 + 16	4 + 3	7 + 7
12 + 1	1 + 8	6 + 5	1 + 5	3 + 14	8 + 7
14 + 2	5 + 9	6 + 1	9 + 8	4 + 7	10 + 8

MATH MAZE

Help the ginger man get out of the maze. Make a path
by drawing a line through the boxes that have sum of 100.

32 + 68	15 + 85	73 + 27	22 + 34	43 + 57	35 + 40	11 + 89
16 + 84	16 + 62	42 + 58	38 + 61	3 + 97	19 + 58	25 + 75	6 + 33

32 + 68	15 + 85	73 + 27	22 + 34	43 + 57	35 + 40	11 + 89	
16 + 84	16 + 62	42 + 58	38 + 61	3 + 97	19 + 58	25 + 75	6 + 33
26 + 74	13 + 82	25 + 75 100			36 + 31	4 + 96	24 + 17
15 + 85	16 + 43	18 + 57			2 + 73	17 + 83	17 + 83
23 + 77	42 + 58	38 + 20	14 + 43	7 + 20	25 + 40	6 + 42	16 + 84
17 + 58	12 + 88	31 + 44	36 + 64	4 + 96	24 + 76	43 + 22	57 + 43
43 + 45	21 + 79	5 + 95	36 + 64	4 + 79	24 + 76	57 + 43	33 + 67

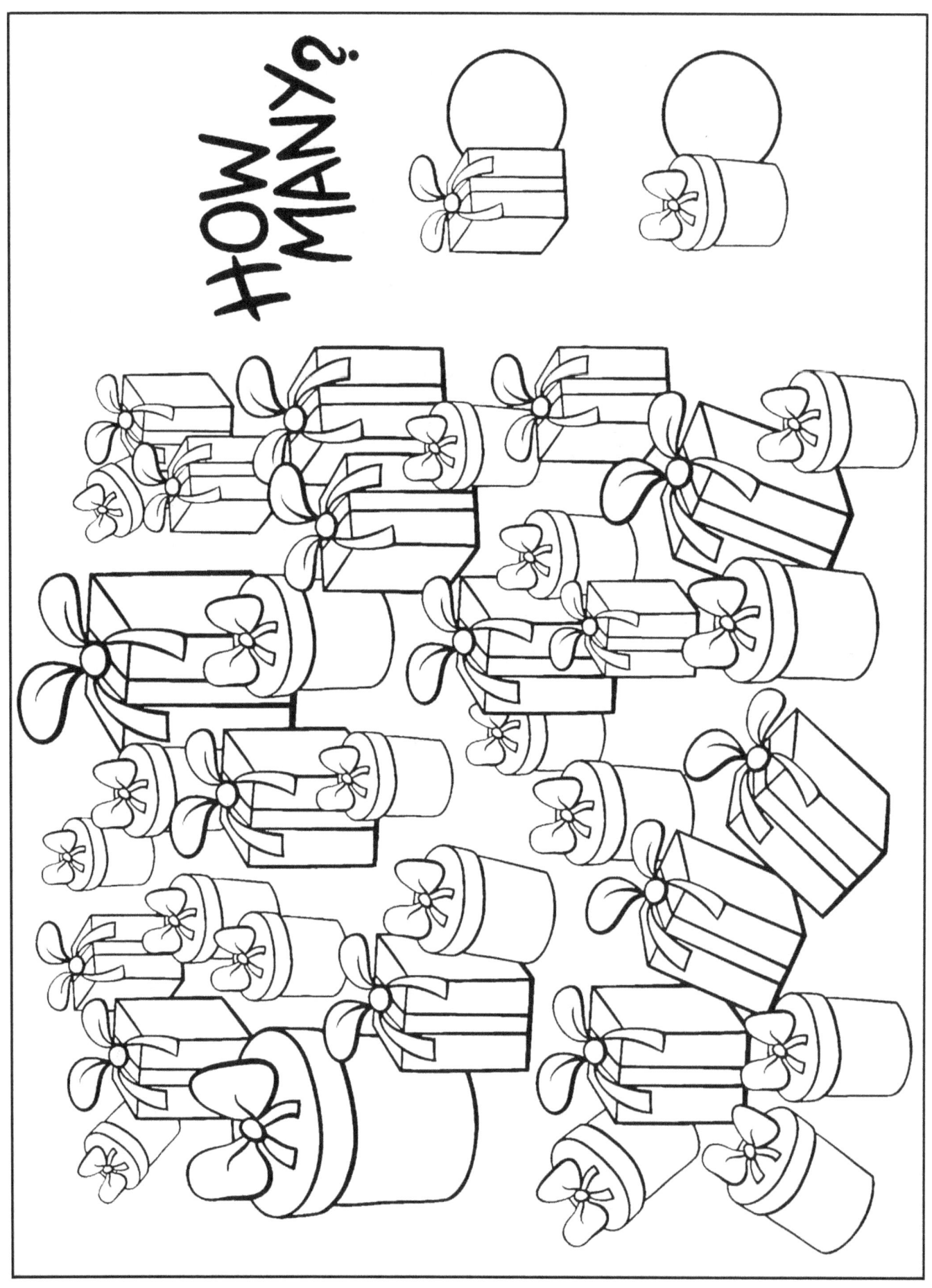

HOW MANY?

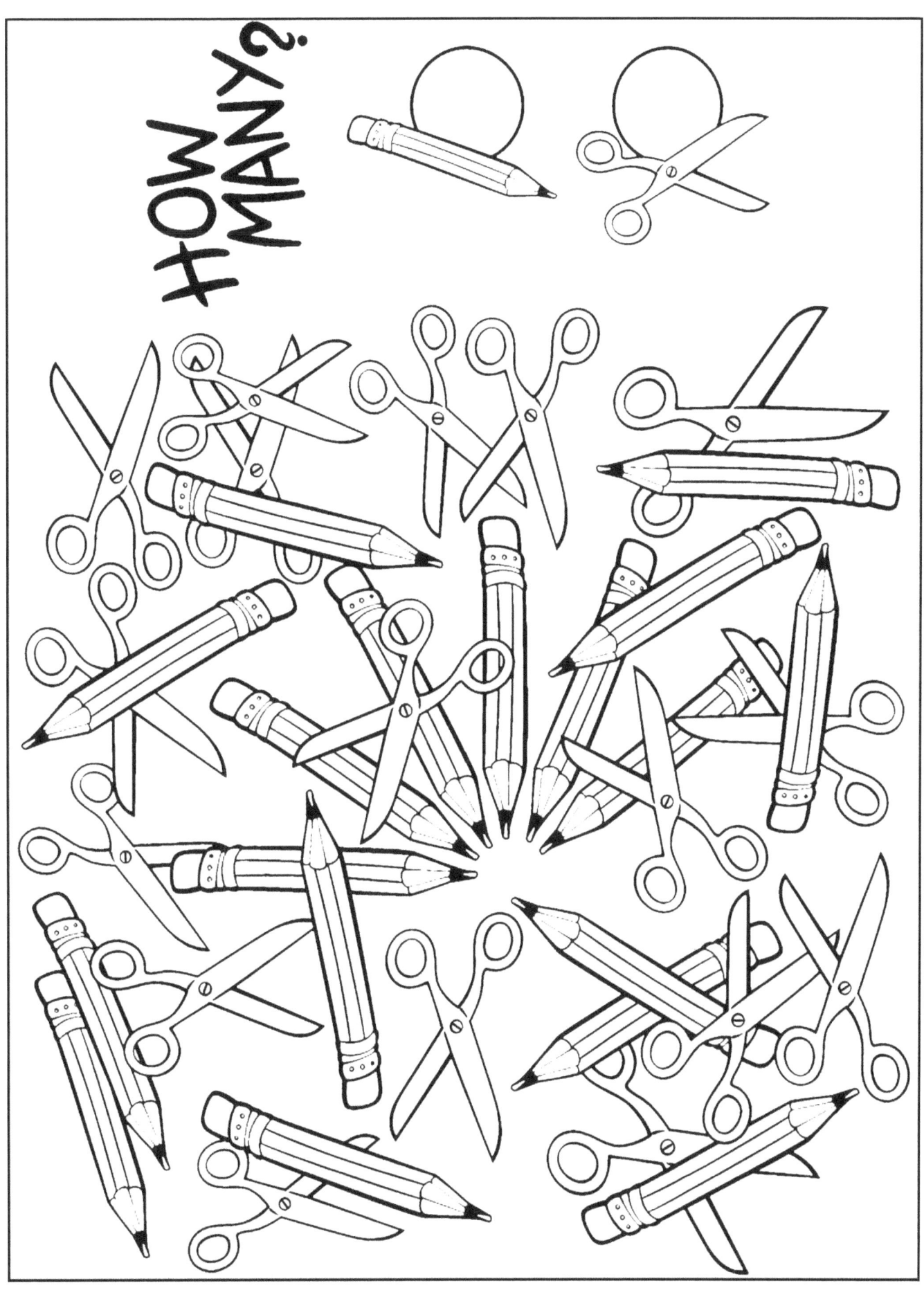

HOW MANY?

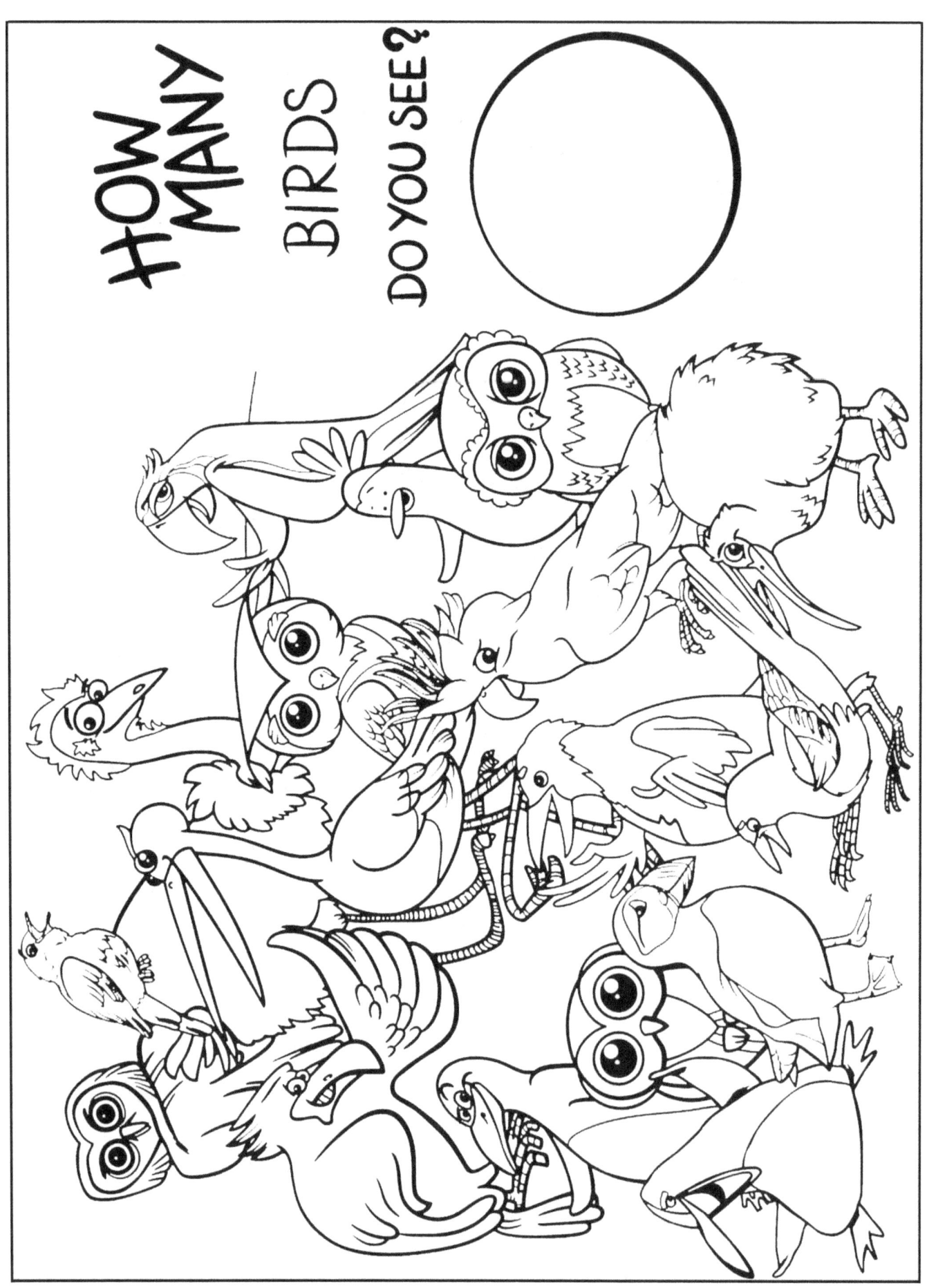

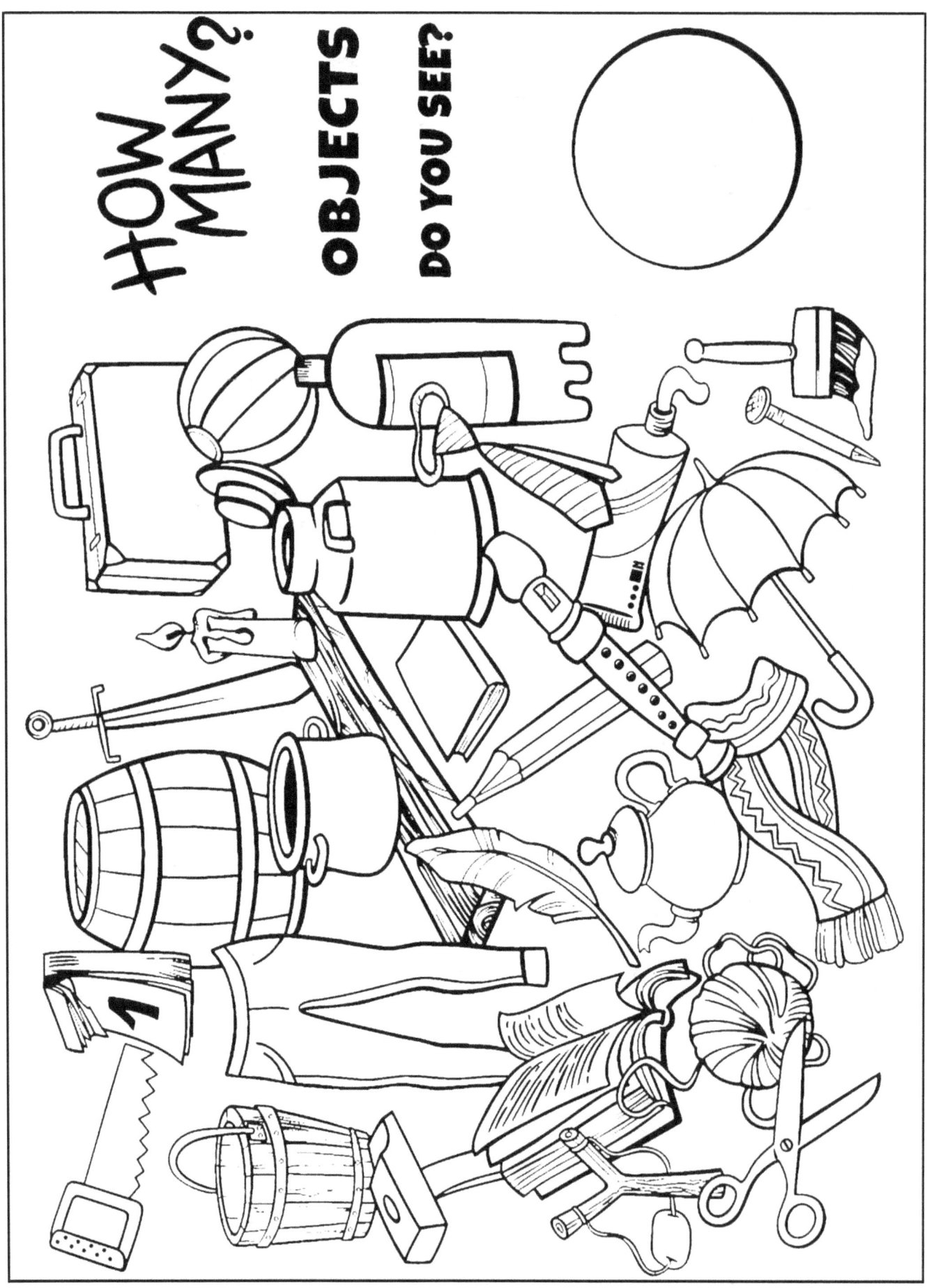

DO YOU SEE?

BULLS

COWS

HOW MANY

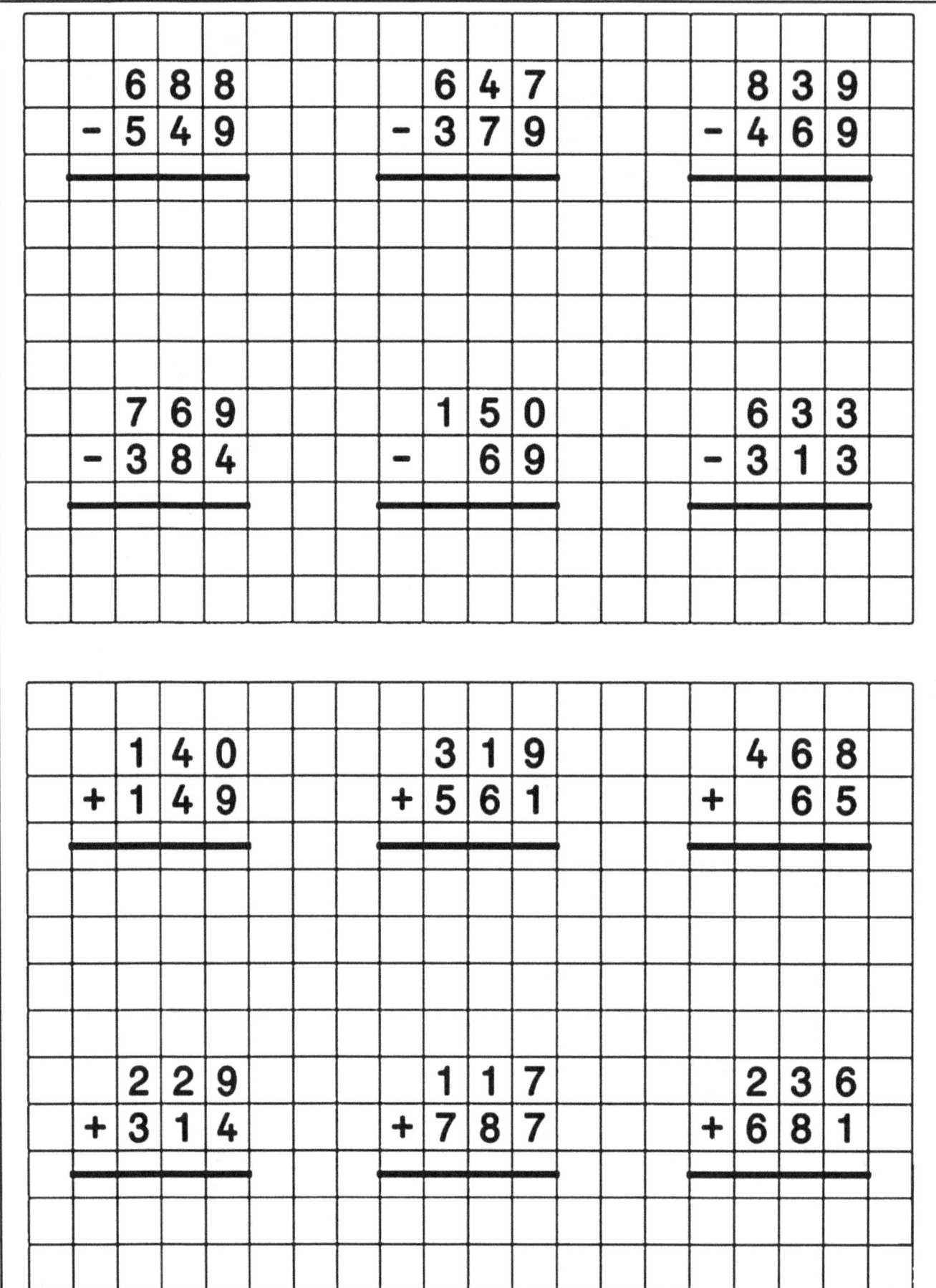

```
  6 8 8        6 4 7        8 3 9
- 5 4 9      - 3 7 9      - 4 6 9
_____      _____      _____



  7 6 9        1 5 0        6 3 3
- 3 8 4      -   6 9      - 3 1 3
_____      _____      _____
```

```
  1 4 0        3 1 9        4 6 8
+ 1 4 9      + 5 6 1      +   6 5
_____      _____      _____



  2 2 9        1 1 7        2 3 6
+ 3 1 4      + 7 8 7      + 6 8 1
_____      _____      _____
```

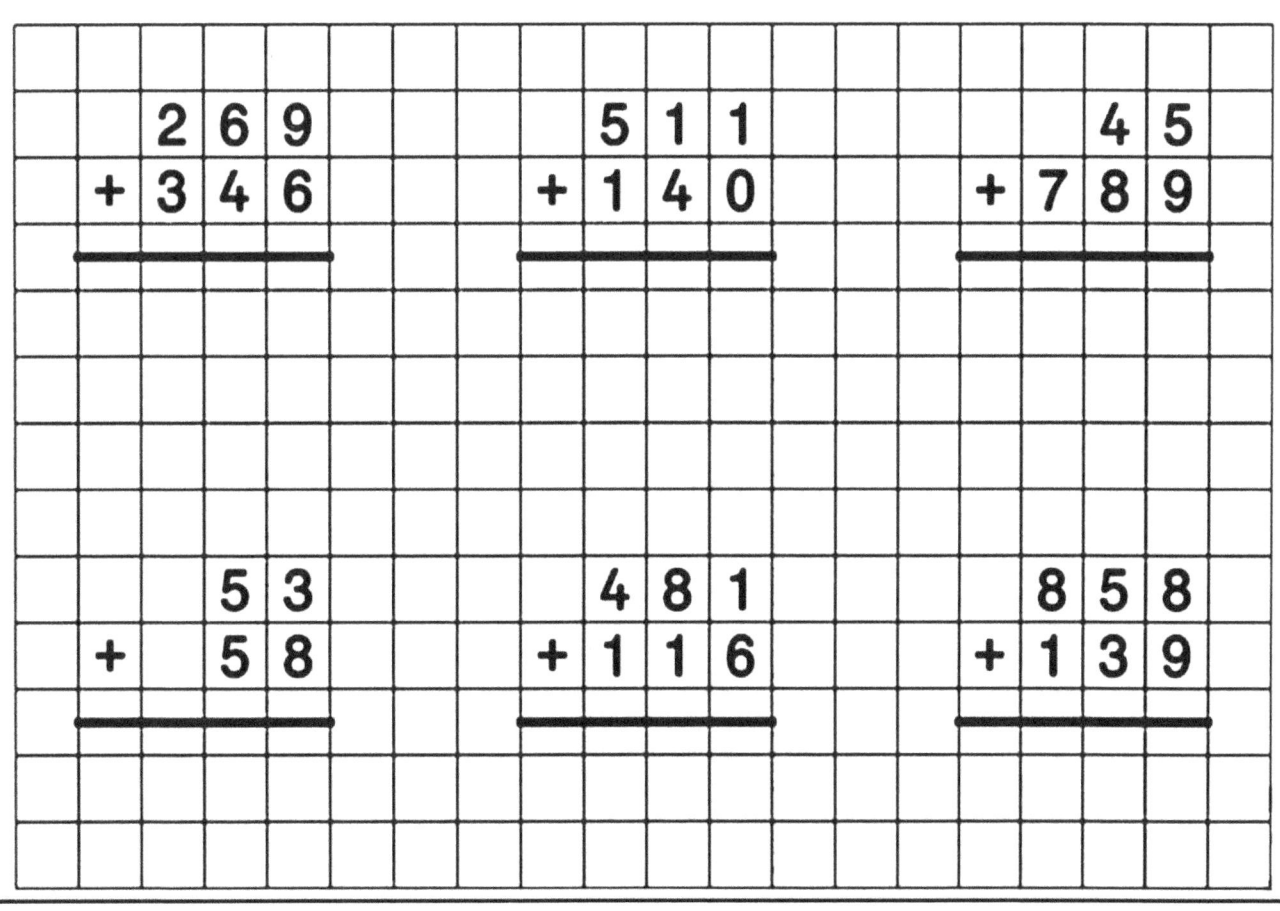

```
   7 3 9          7 2 8          8 1 8
 - 6 3 1        - 1 2 6        - 3 5 6
 _____        _____        _____


   8 3 0          8 3 0
 - 7 2 5        - 6 7 7
 _____        _____
```

```
   2 6 9          5 1 1            4 5
 + 3 4 6        + 1 4 0        + 7 8 9
 _____        _____        _____


     5 3          4 8 1          8 5 8
 +   5 8        + 1 1 6        + 1 3 9
 _____        _____        _____
```

```
  4 0 8          2 6 5          3 0 9
- 2 5 7        - 1 8 3        - 2 8 8
_____        _____       _____


  7 0 8          2 2 6          5 5 5
- 5 8 0        - 1 1 6        - 4 4 8
_____        _____       _____
```

```
    5 3 9            1 2 1
  +   4          + 3 3 0
  _____          _____


    8 8 1            5 0 8
  +   7 4          +   5 9
  _____          _____
```

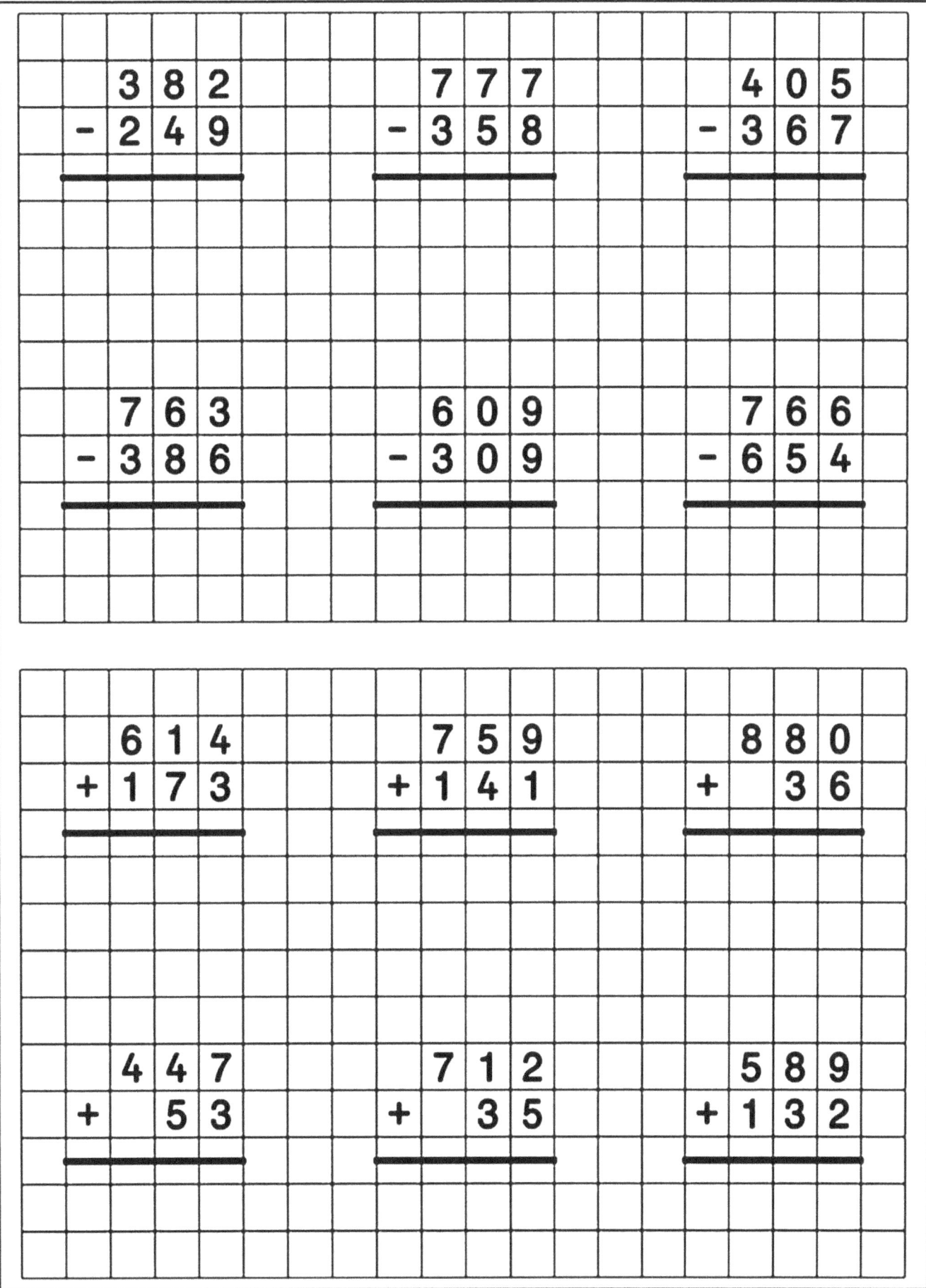

```
  3 8 2         7 7 7         4 0 5
- 2 4 9       - 3 5 8       - 3 6 7
_____       _____       _____



  7 6 3         6 0 9         7 6 6
- 3 8 6       - 3 0 9       - 6 5 4
_____       _____       _____
```

```
  6 1 4         7 5 9         8 8 0
+ 1 7 3       + 1 4 1       +   3 6
_____       _____       _____



  4 4 7         7 1 2         5 8 9
+   5 3       +   3 5       + 1 3 2
_____       _____       _____
```

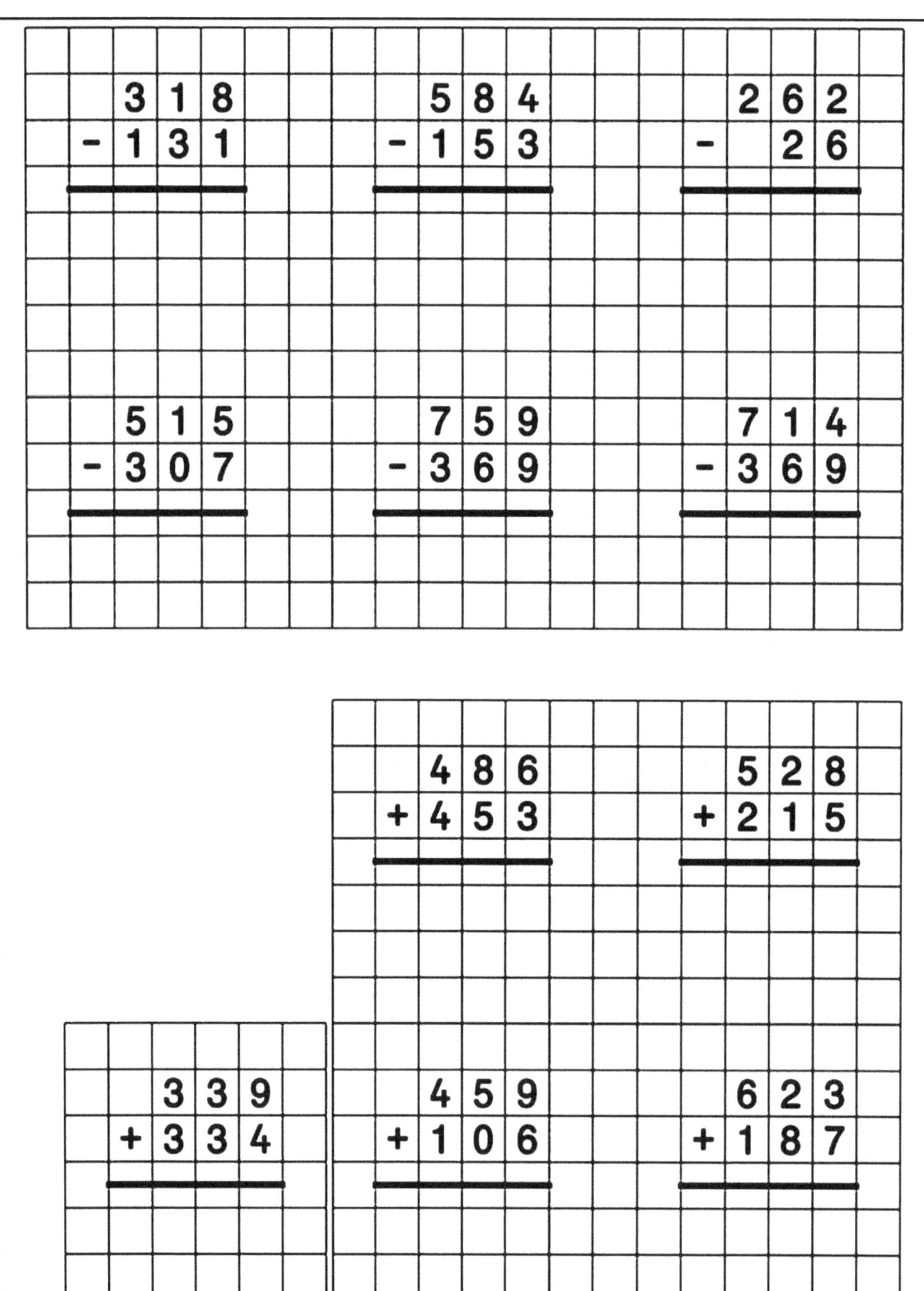

```
    3 1 8          5 8 4          2 6 2
  - 1 3 1        - 1 5 3        -   2 6
  ─────────      ─────────      ─────────



    5 1 5          7 5 9          7 1 4
  - 3 0 7        - 3 6 9        - 3 6 9
  ─────────      ─────────      ─────────
```

```
                   4 8 6          5 2 8
                 + 4 5 3        + 2 1 5
                 ─────────      ─────────



    3 3 9          4 5 9          6 2 3
  + 3 3 4        + 1 0 6        + 1 8 7
  ─────────      ─────────      ─────────
```

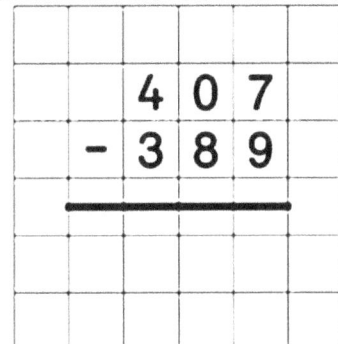

$$
\begin{array}{r}
4\ 0\ 7 \\
-\ 3\ 8\ 9 \\
\hline
\end{array}
$$

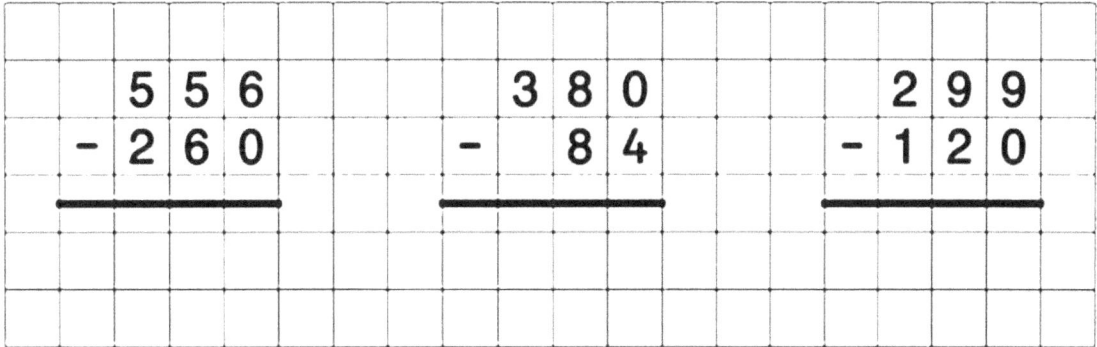

$$
\begin{array}{r}
5\ 5\ 6 \\
-\ 2\ 6\ 0 \\
\hline
\end{array}
\qquad
\begin{array}{r}
3\ 8\ 0 \\
-\ \ \ 8\ 4 \\
\hline
\end{array}
\qquad
\begin{array}{r}
2\ 9\ 9 \\
-\ 1\ 2\ 0 \\
\hline
\end{array}
$$

5 · 7 = ____   4 · 7 = ____   2 · 4 = ____

5 · 1 = ____   4 · 1 = ____   2 · 1 = ____

5 · 6 = ____   4 · 8 = ____   2 · 5 = ____

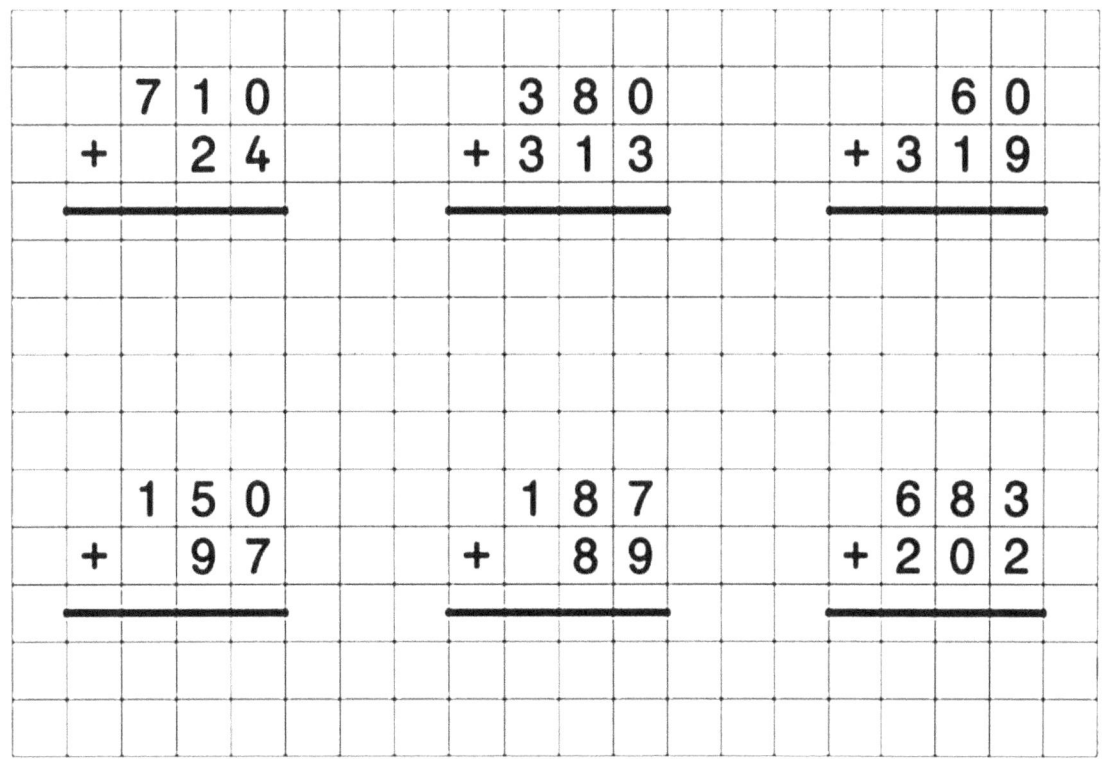

$$
\begin{array}{r}
7\ 1\ 0 \\
+\ \ \ 2\ 4 \\
\hline
\end{array}
\qquad
\begin{array}{r}
3\ 8\ 0 \\
+\ 3\ 1\ 3 \\
\hline
\end{array}
\qquad
\begin{array}{r}
\ \ \ 6\ 0 \\
+\ 3\ 1\ 9 \\
\hline
\end{array}
$$

$$
\begin{array}{r}
1\ 5\ 0 \\
+\ \ \ 9\ 7 \\
\hline
\end{array}
\qquad
\begin{array}{r}
1\ 8\ 7 \\
+\ \ \ 8\ 9 \\
\hline
\end{array}
\qquad
\begin{array}{r}
6\ 8\ 3 \\
+\ 2\ 0\ 2 \\
\hline
\end{array}
$$

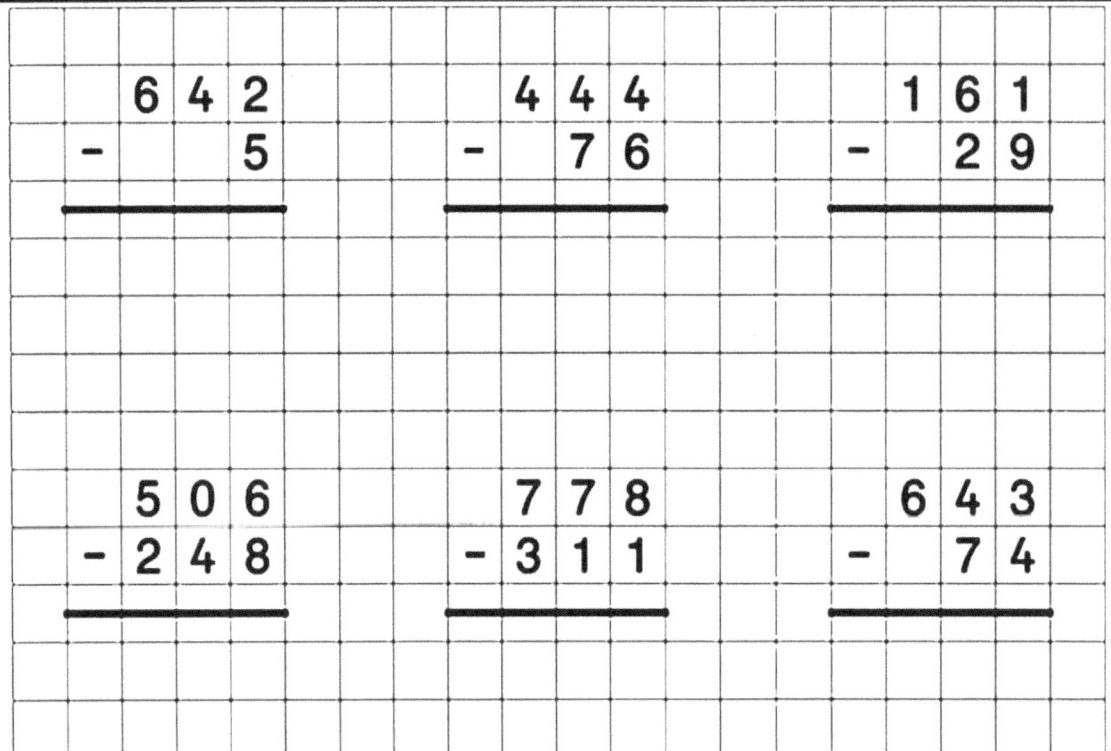

```
    6 4 2          4 4 4          1 6 1
  -     5        -   7 6        -   2 9
  _____    _____    _____


    5 0 6          7 7 8          6 4 3
  - 2 4 8        - 3 1 1        -   7 4
  _____    _____    _____
```

$5 \cdot 9 = \underline{\hspace{1.5em}}$ $4 \cdot 9 = \underline{\hspace{1.5em}}$ $2 \cdot 2 = \underline{\hspace{1.5em}}$

$5 \cdot 4 = \underline{\hspace{1.5em}}$ $4 \cdot 6 = \underline{\hspace{1.5em}}$ $2 \cdot 6 = \underline{\hspace{1.5em}}$

$5 \cdot 5 = \underline{\hspace{1.5em}}$ $4 \cdot 8 = \underline{\hspace{1.5em}}$ $2 \cdot 3 = \underline{\hspace{1.5em}}$

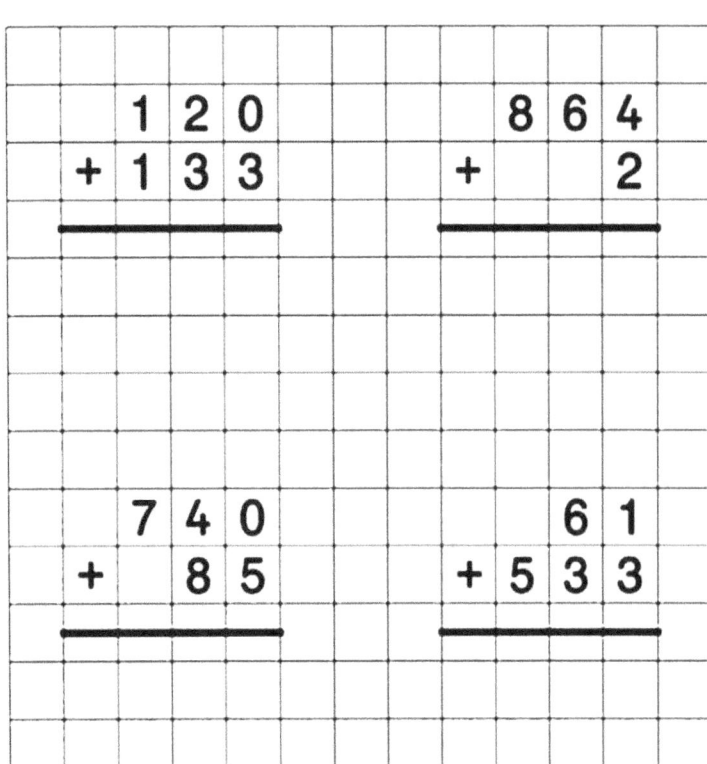

```
    1 2 0          8 6 4
  + 1 3 3        +     2
  _____    _____


    7 4 0            6 1
  +   8 5        + 5 3 3
  _____    _____
```

```
    3 4 0              5 8 8              2 6 3
  - 2 3 3            - 2 4 6            - 1 6 7
  ─────────          ─────────          ─────────


    2 0 3              2 7 2
  -   5 7            - 1 6 1
  ─────────          ─────────
```

5 · 1 = ___   4 · 1 = ___   3 · 3 = ___

5 · 4 = ___   4 · 9 = ___   3 · 2 = ___

5 · 5 = ___   4 · 4 = ___   3 · 7 = ___

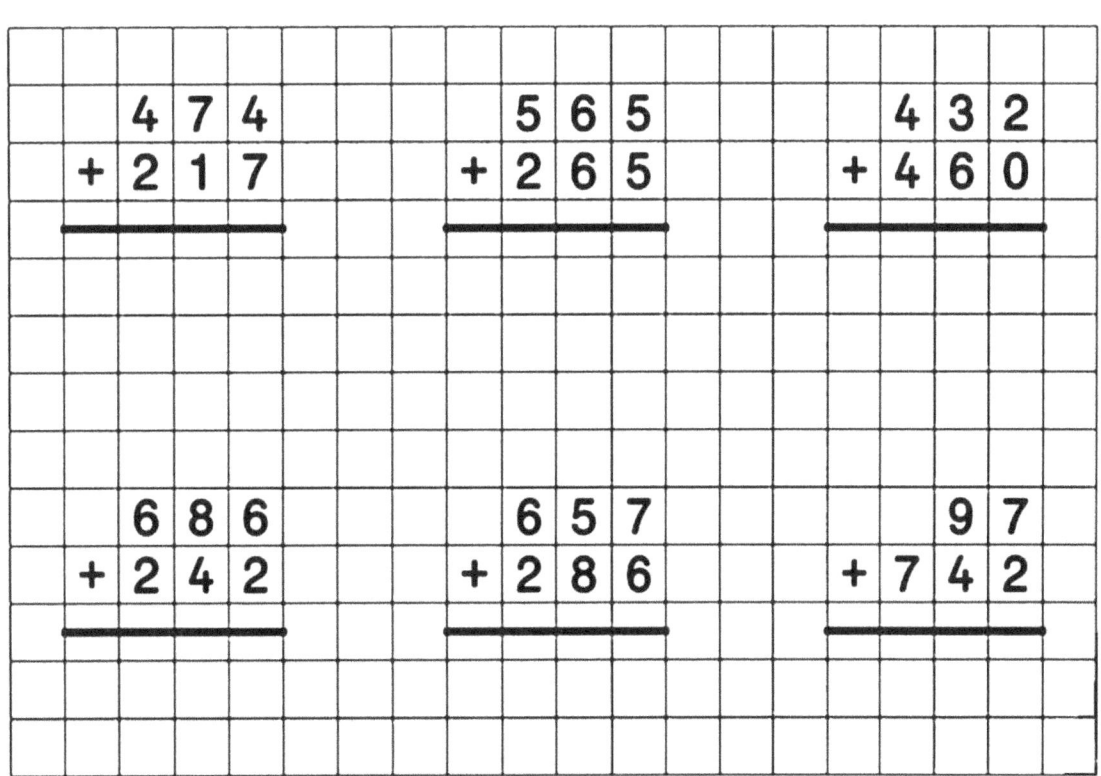

```
    4 7 4              5 6 5              4 3 2
  + 2 1 7            + 2 6 5            + 4 6 0
  ─────────          ─────────          ─────────




    6 8 6              6 5 7                9 7
  + 2 4 2            + 2 8 6            + 7 4 2
  ─────────          ─────────          ─────────
```

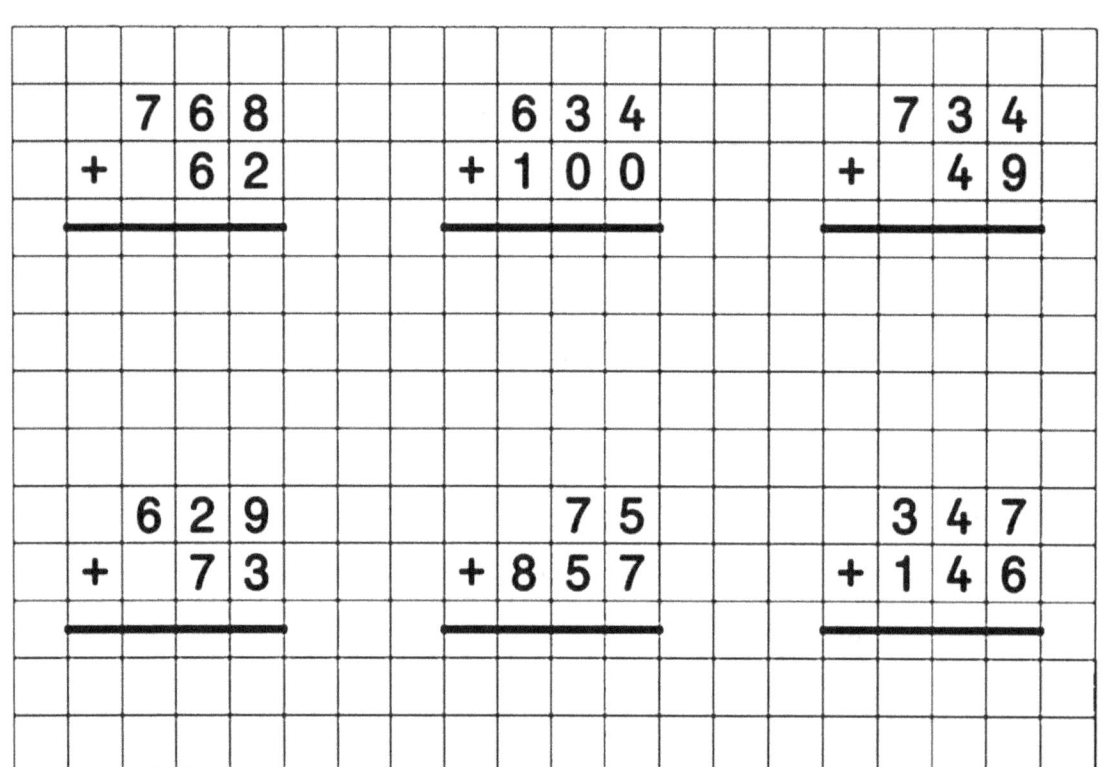

```
  8 8 3          4 2 0          2 4 1
- 5 4 7        - 3 0 6        - 1 5 7
─────────      ─────────      ─────────


  7 6 7          1 8 8          1 3 0
- 2 6 5        -   8 1        -   1 0
─────────      ─────────      ─────────
```

5 · 9 = _____   4 · 5 = _____   3 · 4 = _____

5 · 7 = _____   4 · 2 = _____   3 · 6 = _____

5 · 6 = _____   4 · 4 = _____   3 · 3 = _____

```
  7 6 8          6 3 4          7 3 4
+   6 2        + 1 0 0        +   4 9
─────────      ─────────      ─────────


  6 2 9            7 5          3 4 7
+   7 3        + 8 5 7        + 1 4 6
─────────      ─────────      ─────────
```

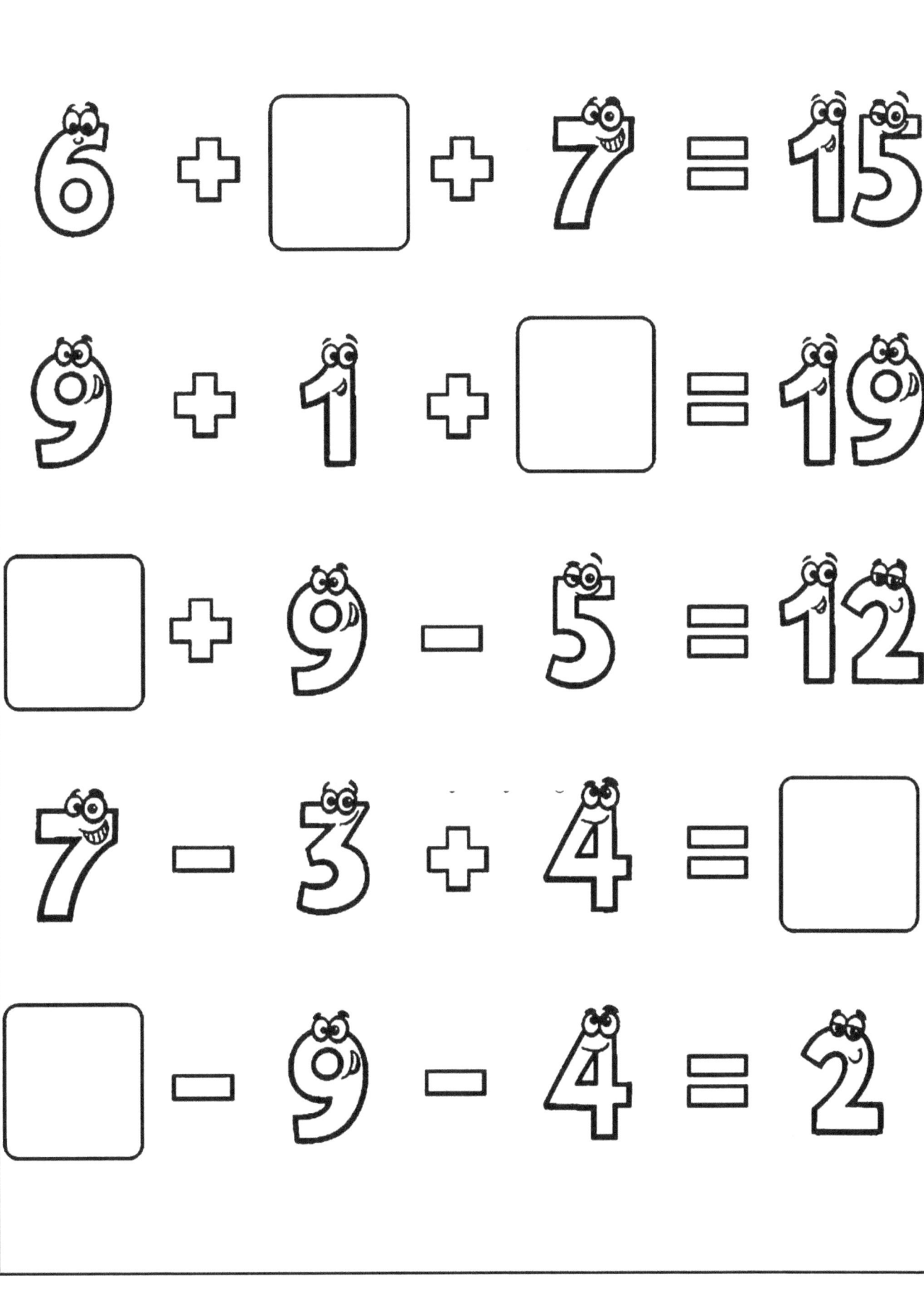

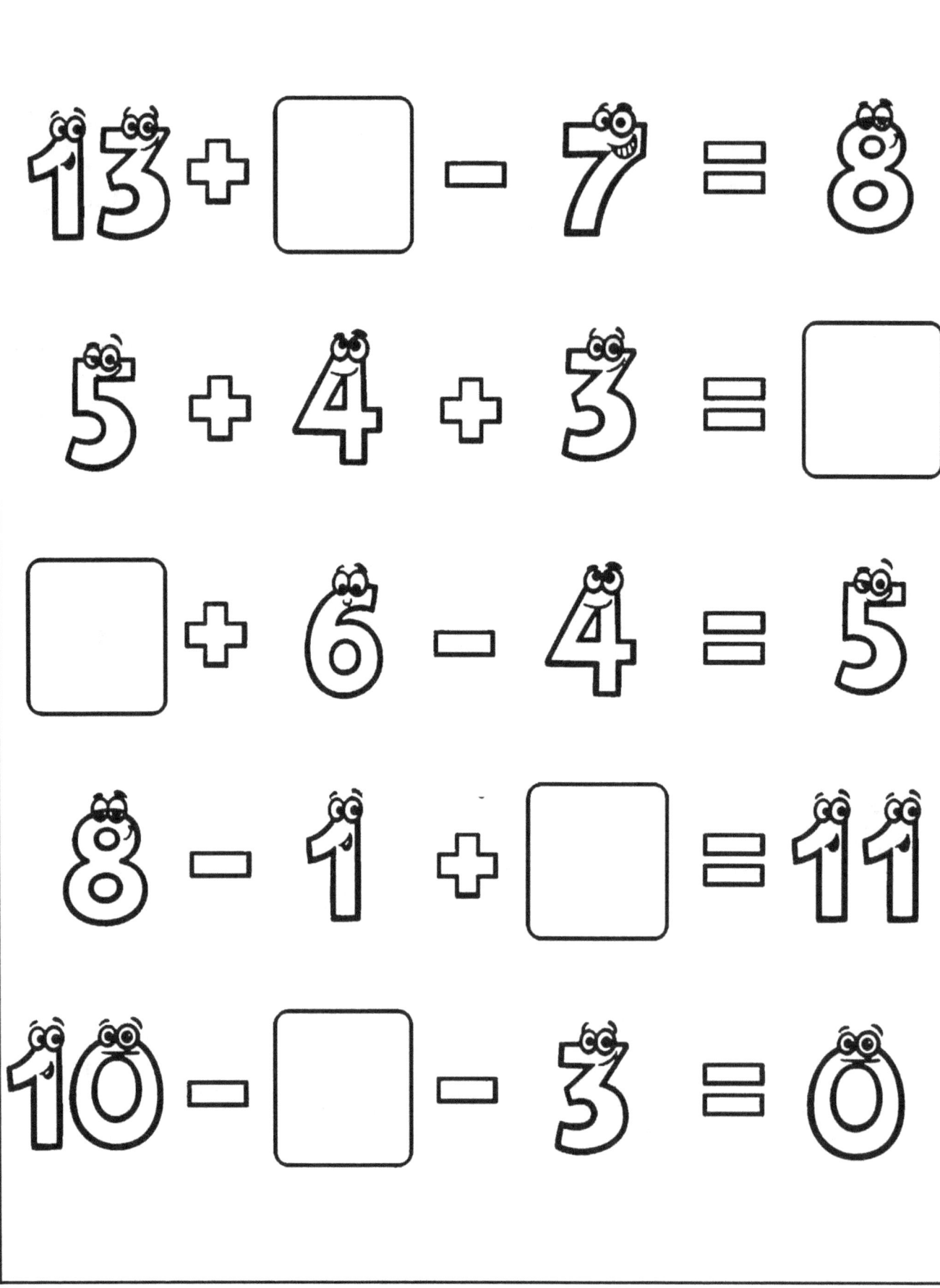

1. ◯ + ⬤6 = ⬤7

2. ⬤10 − ◯ = ⬤9

3. ⬤4 − ◯ = ⬤2

4. ◯ + ⬤3 = ⬤5

5. ◯ + ⬤4 = ⬤14

6. ◯ − ⬤1 = ⬤8

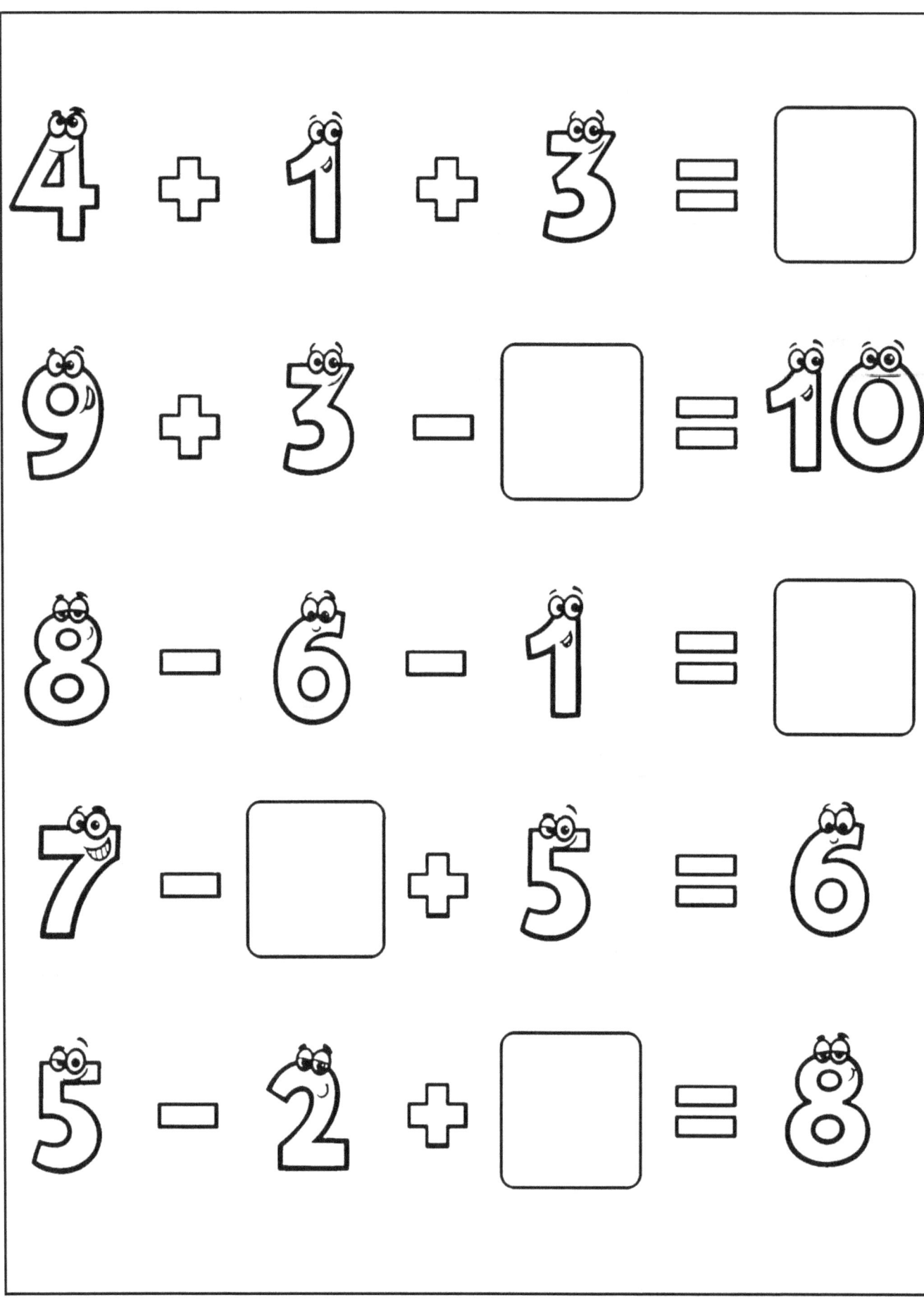

$4 + 1 + 3 = \boxed{}$

$9 + 3 - \boxed{} = 10$

$8 - 6 - 1 = \boxed{}$

$7 - \boxed{} + 5 = 6$

$5 - 2 + \boxed{} = 8$

$$12 - 9 + 2 = \boxed{}$$

$$3 + \boxed{} + 4 = 8$$

$$\boxed{} - 5 - 5 = 10$$

$$7 - 6 + \boxed{} = 3$$

$$4 - \boxed{} + 8 = 9$$

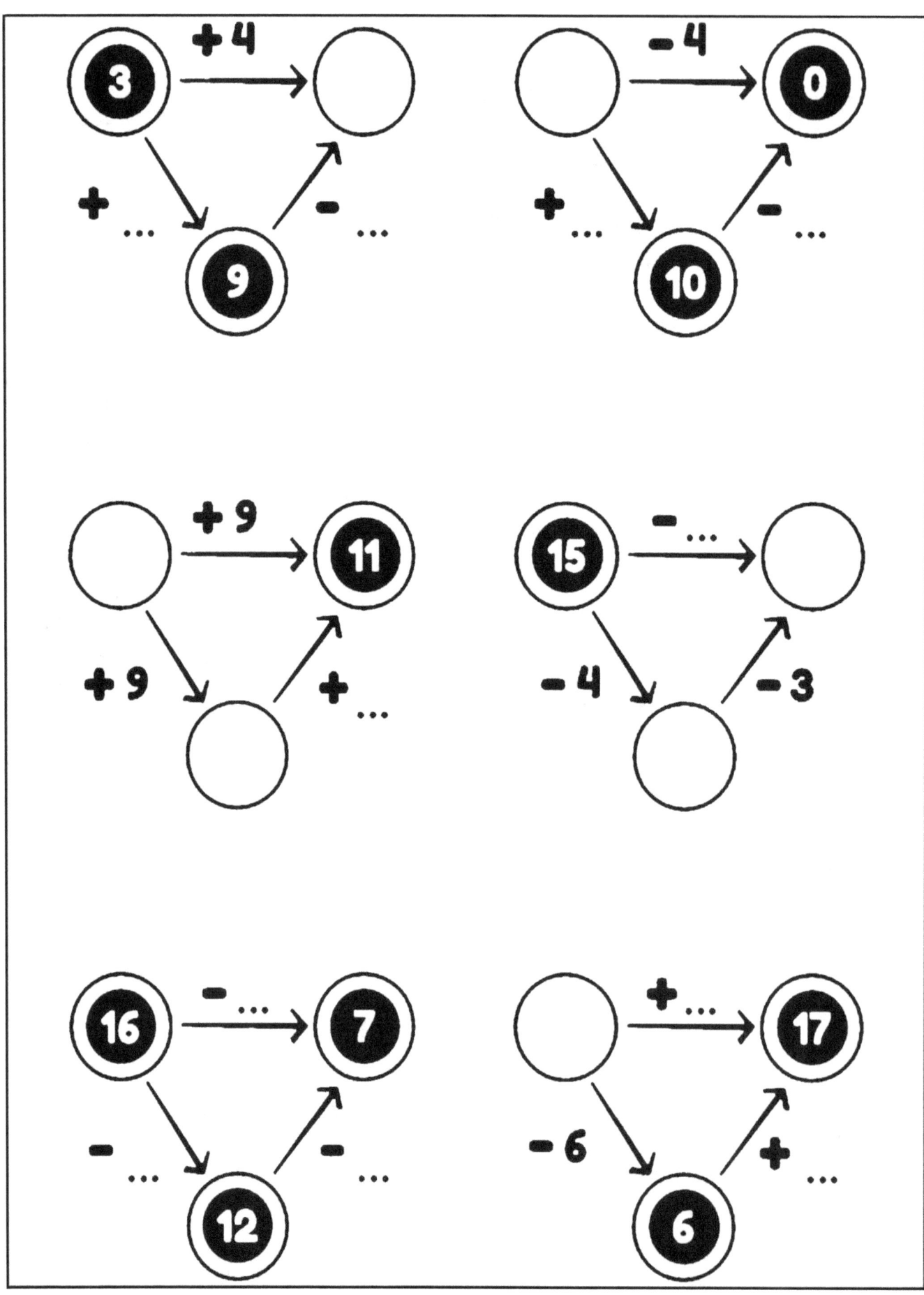

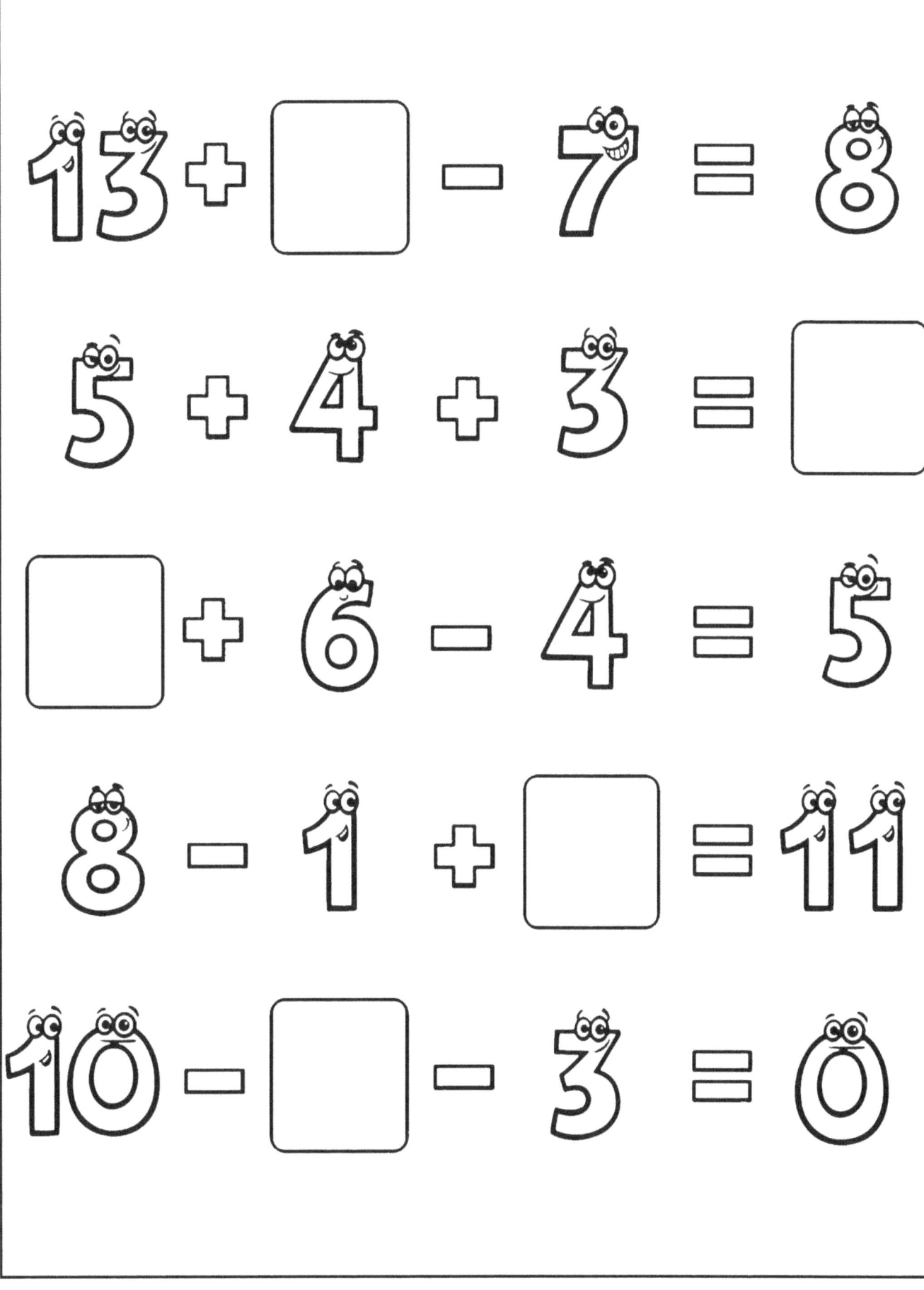

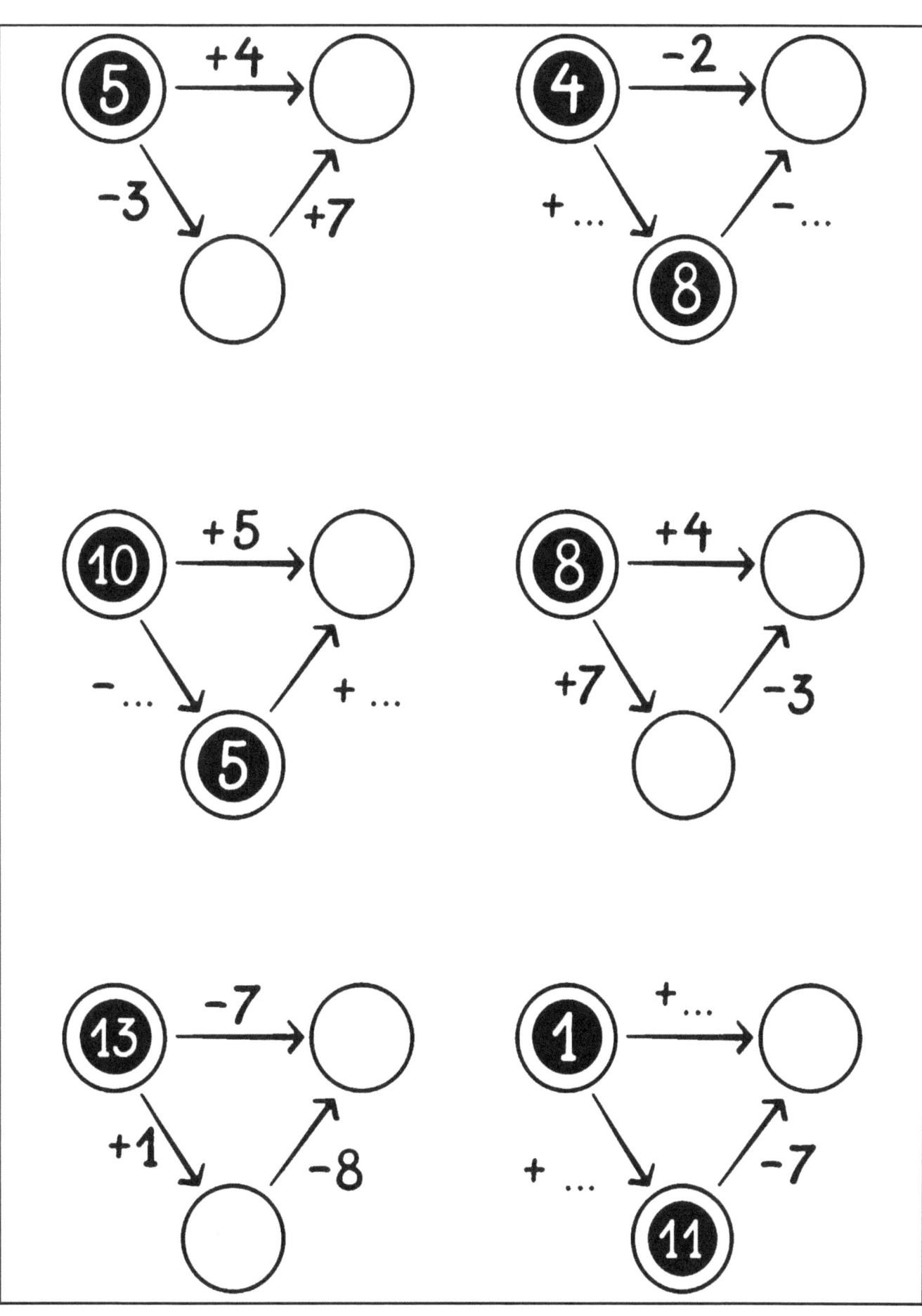

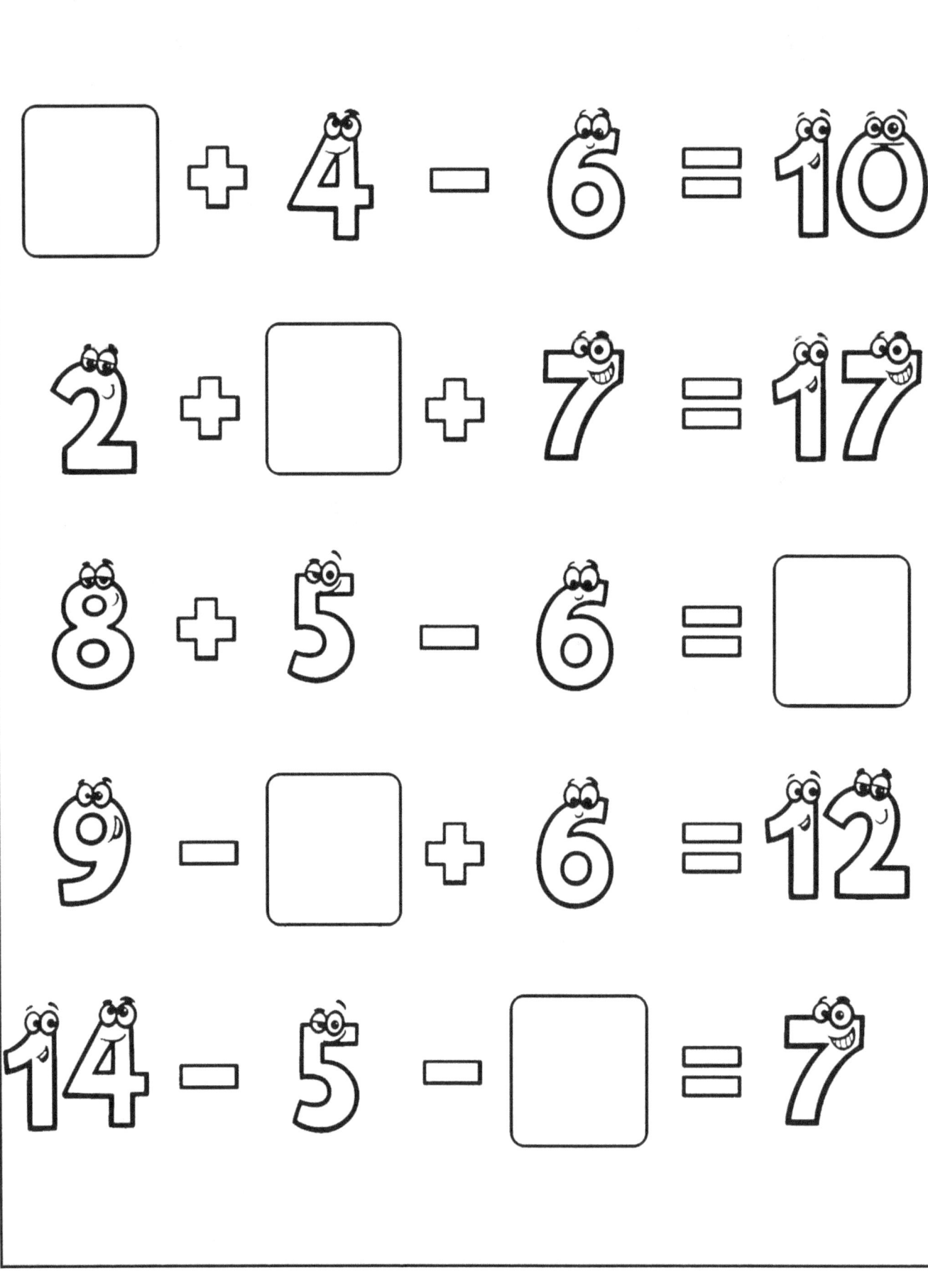

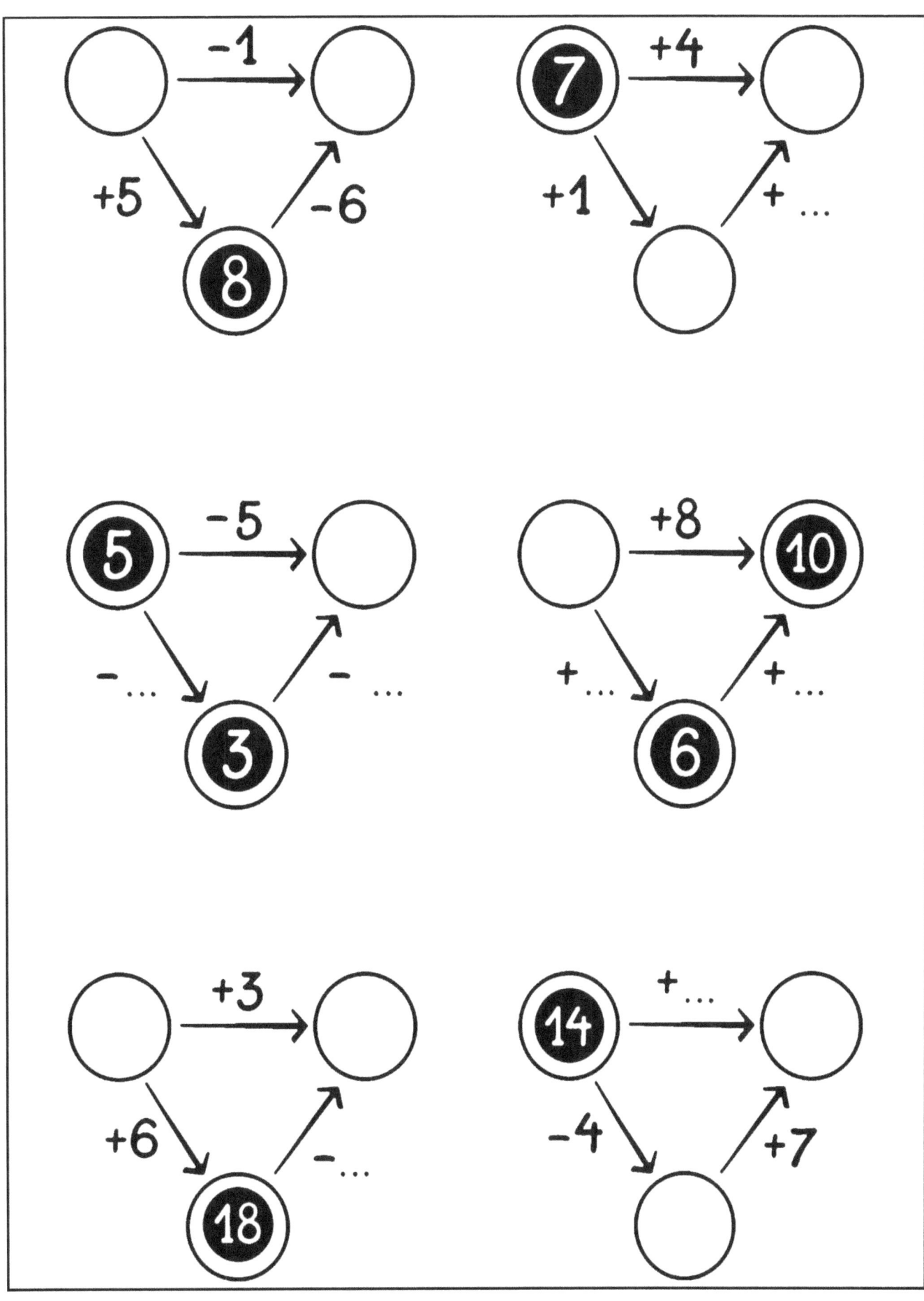

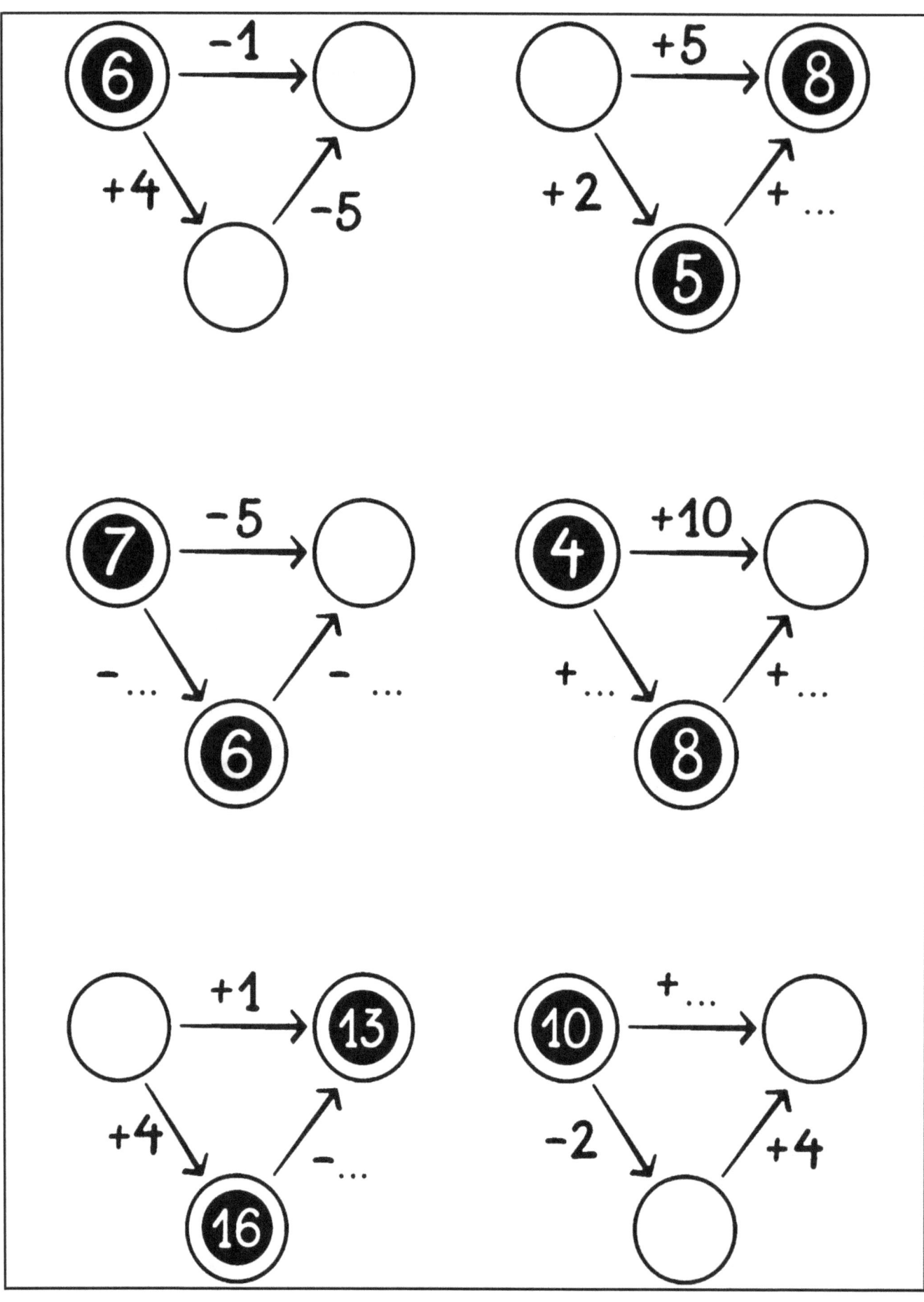

1. ② + ◯ = ⑤

2. ◯ + ③ = ⑧

3. ◯ + ② = ④

4. ◯ + ① = ⑥

5. ⑥ + ◯ = ⑫

6. ◯ - ③ = ⑦

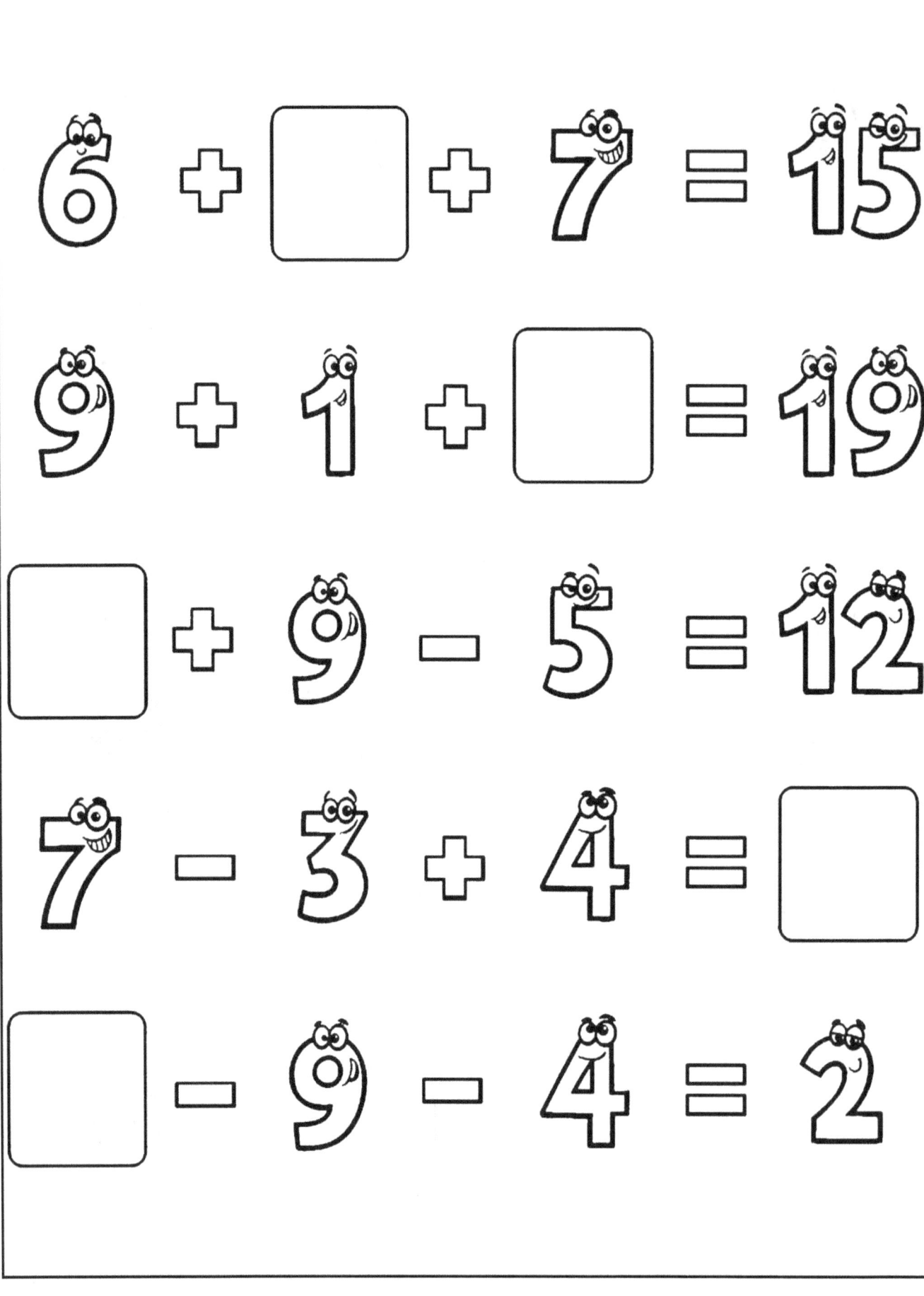

$$5 + \boxed{} + 2 = 12$$

$$3 + 14 - 8 = \boxed{}$$

$$\boxed{} - 9 + 4 = 7$$

$$7 + 11 - \boxed{} = 10$$

$$15 - \boxed{} - 6 = 3$$

1. ◯ + ① = ④

2. ⑩ − ◯ = ⑧

3. ◯ − ⑦ = ②

4. ⑨ + ◯ = ⑫

5. ◯ − ④ = ③

6. ⑪ − ◯ = ⑤

Solve the task

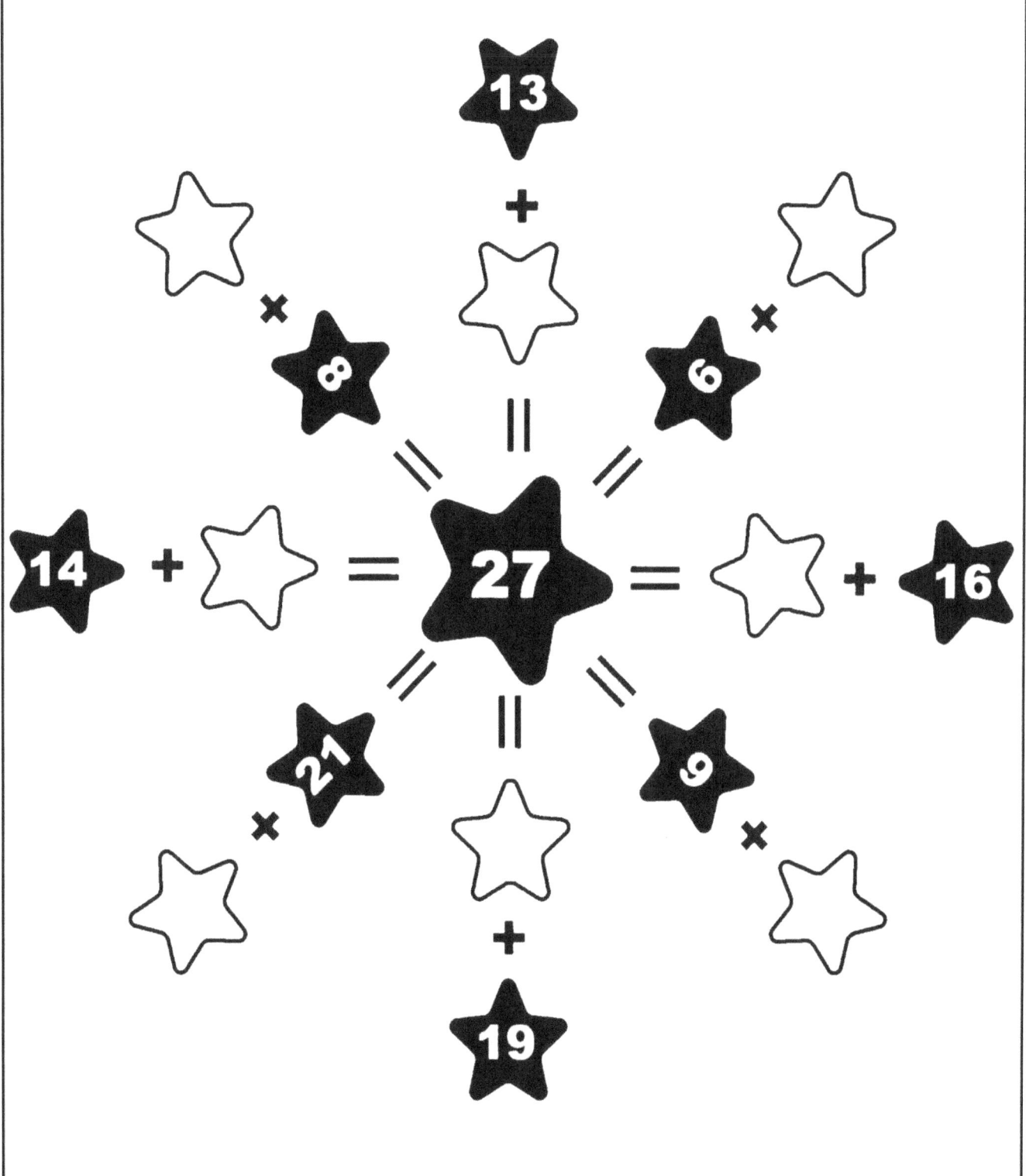

Test Your Color

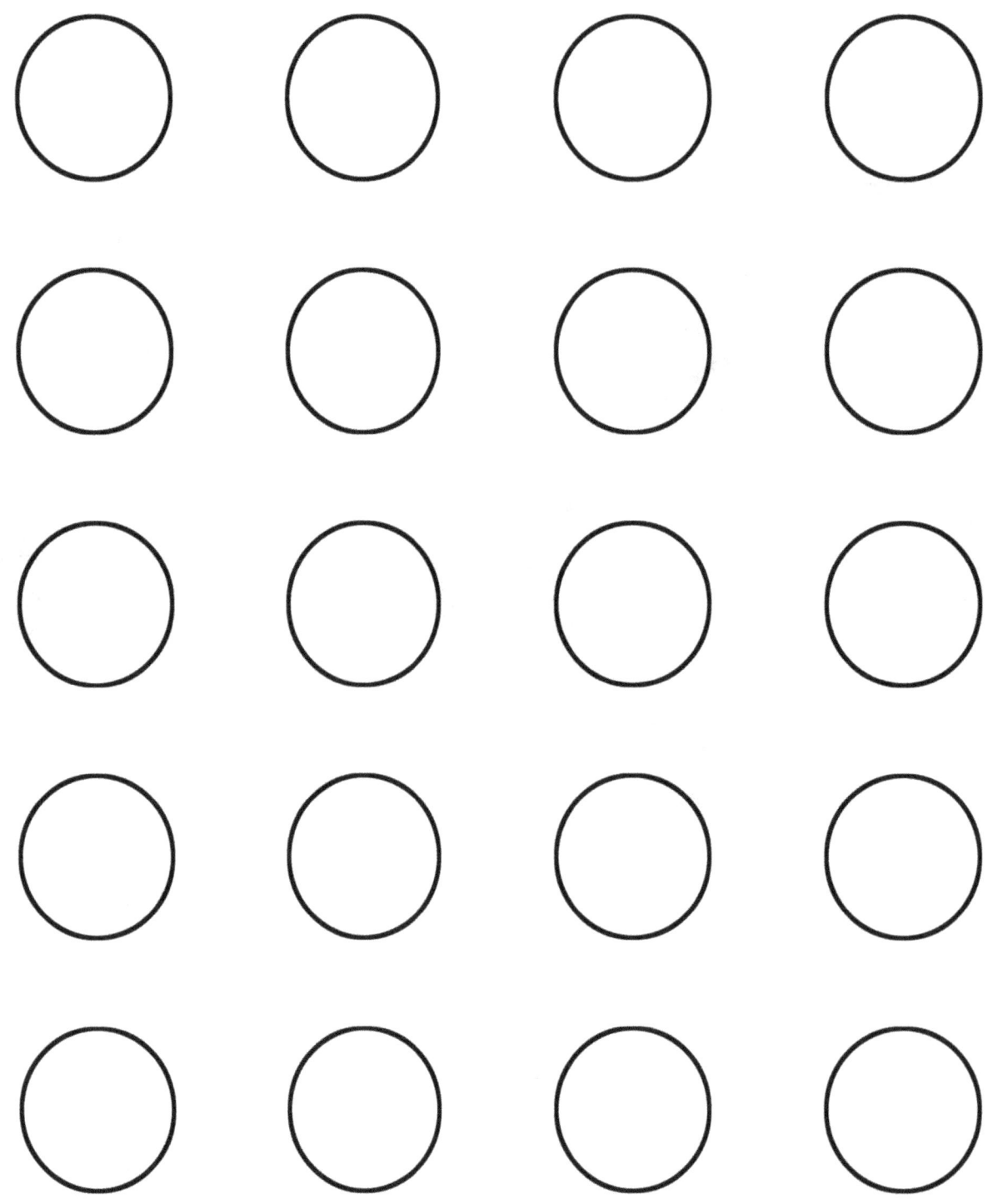

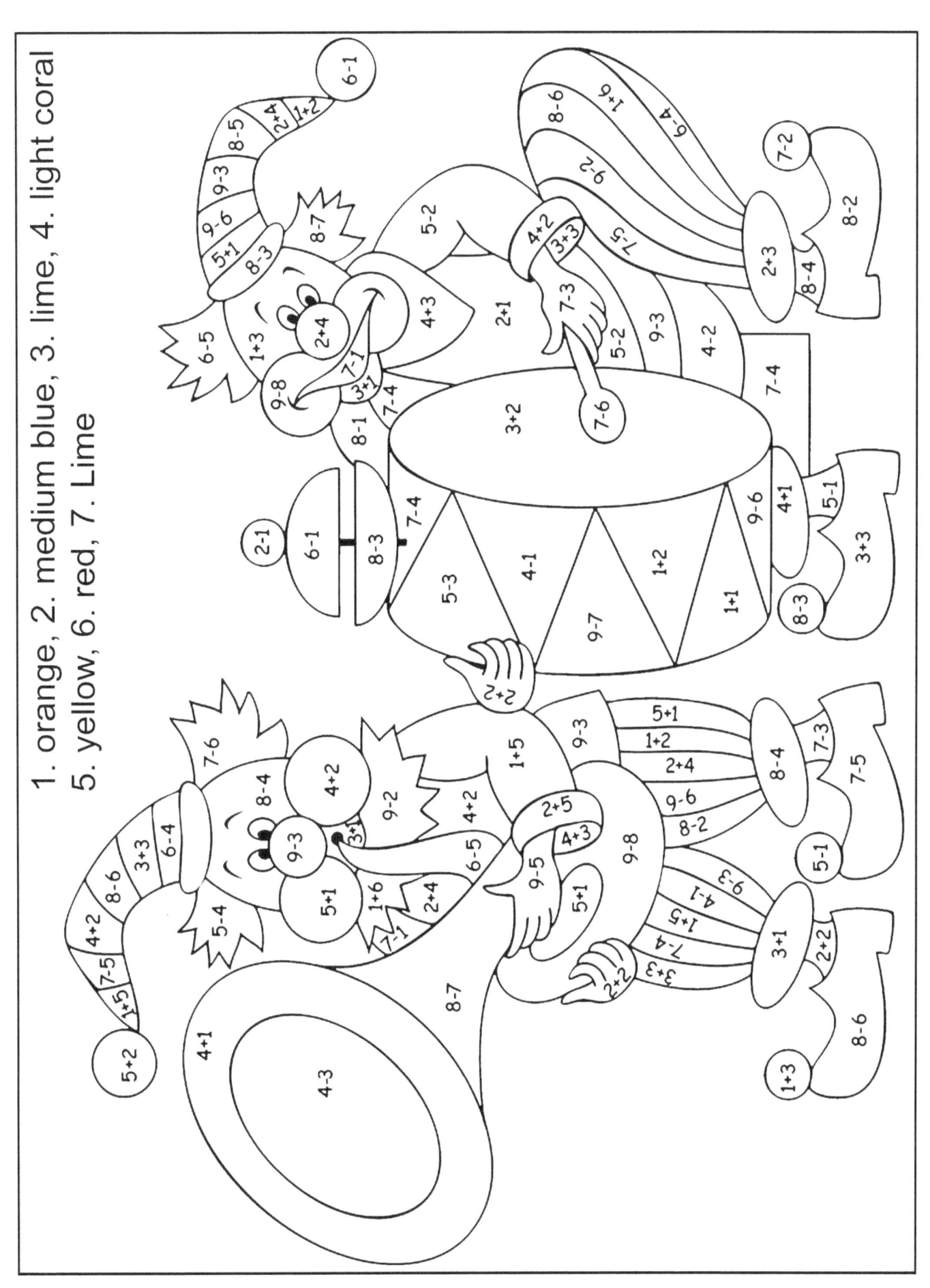

1. orange, 2. medium blue, 3. lime, 4. light coral
5. yellow, 6. red, 7. Lime

Test Your Color

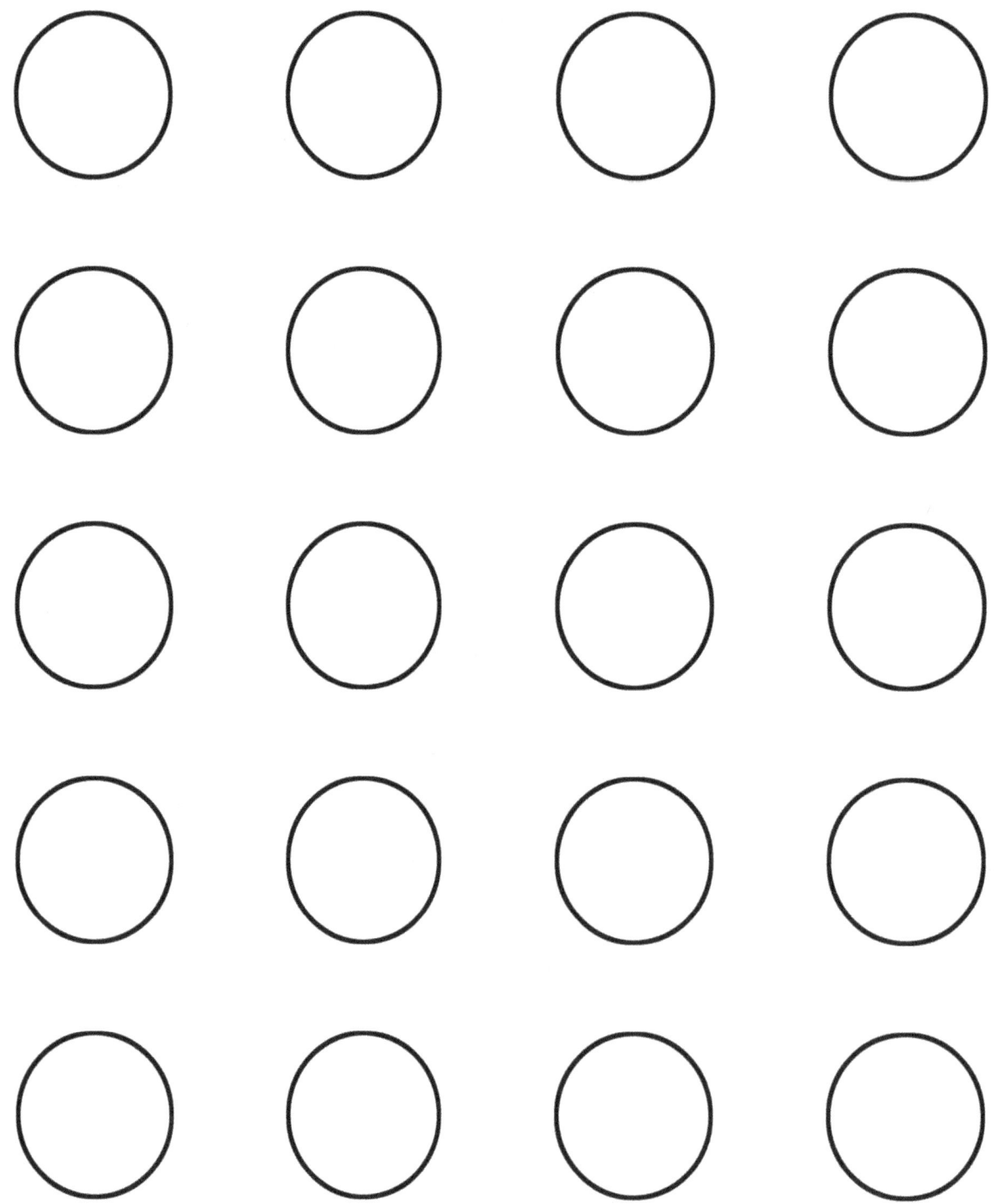

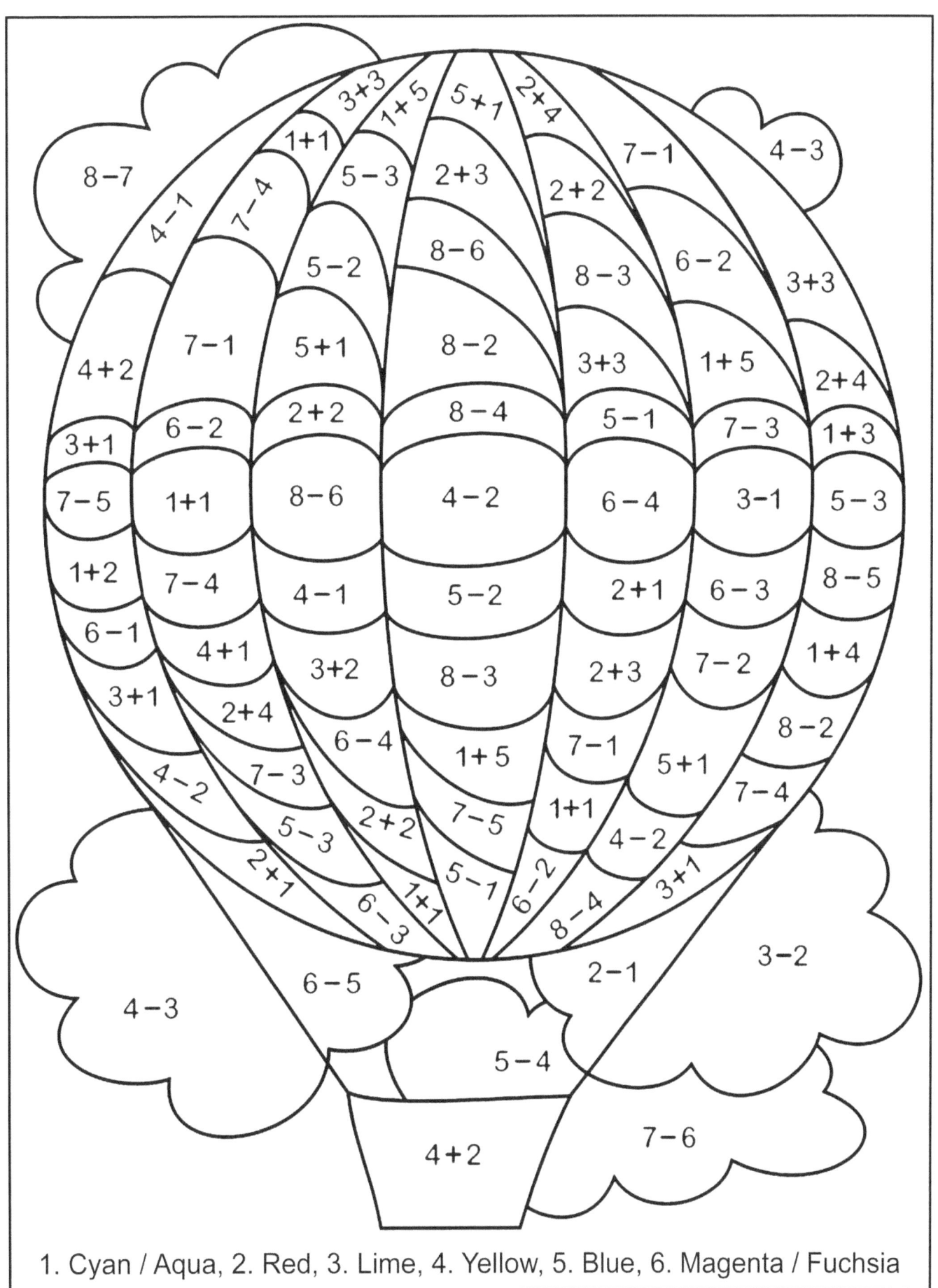

1. Cyan / Aqua, 2. Red, 3. Lime, 4. Yellow, 5. Blue, 6. Magenta / Fuchsia

Test Your Color

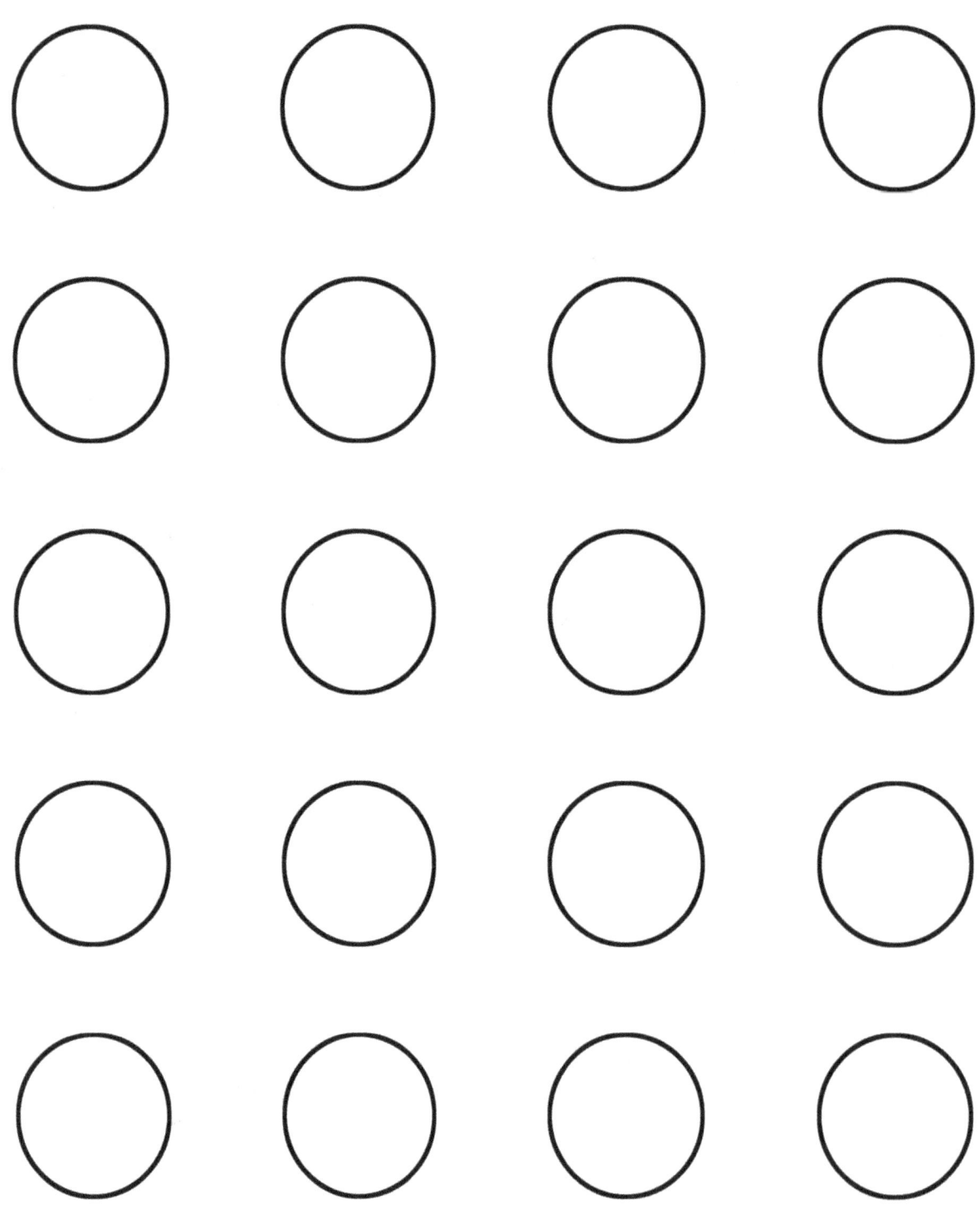

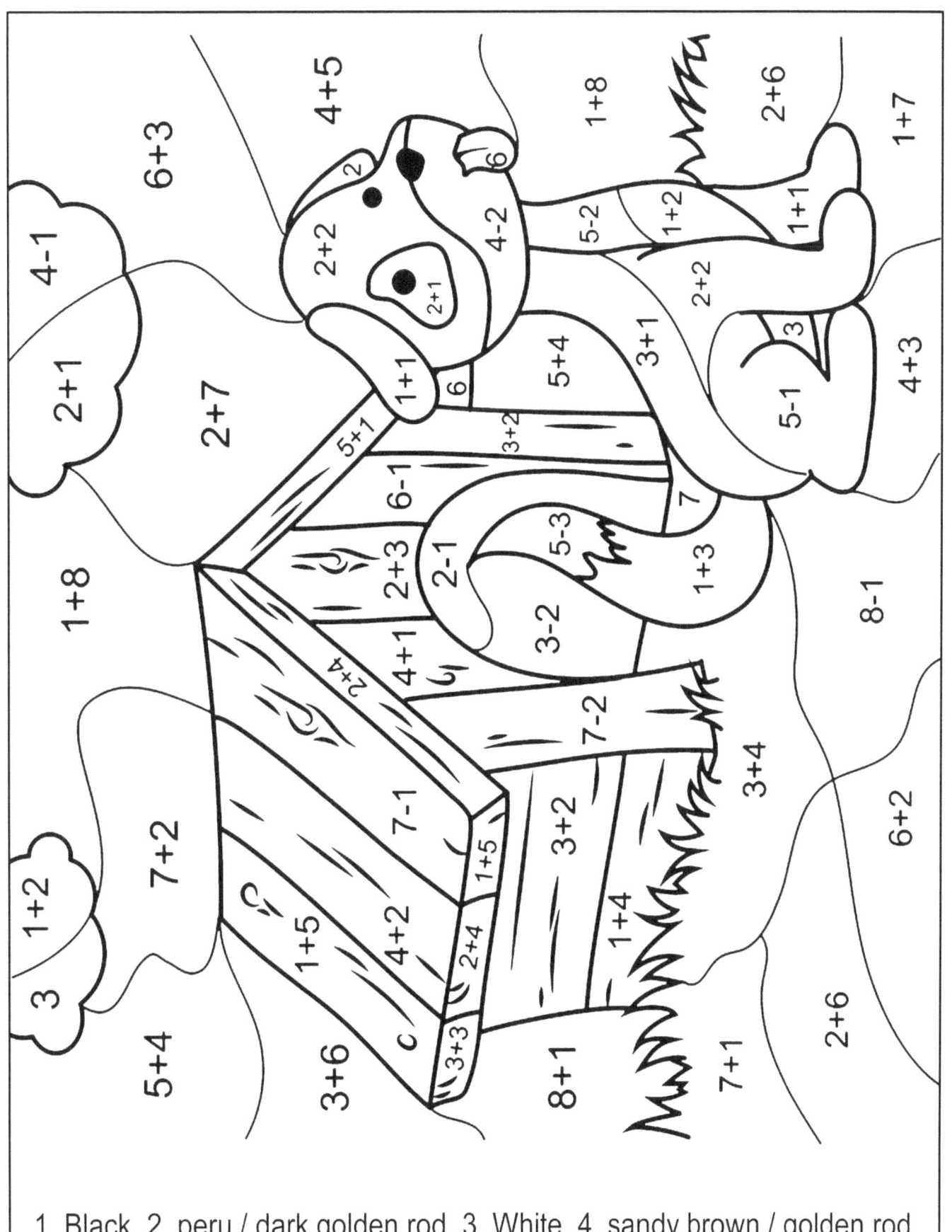

1. Black, 2. peru / dark golden rod, 3. White, 4. sandy brown / golden rod
5. yellow, 6. orange red, 7. lawn green, 8. aqua

Test Your Color

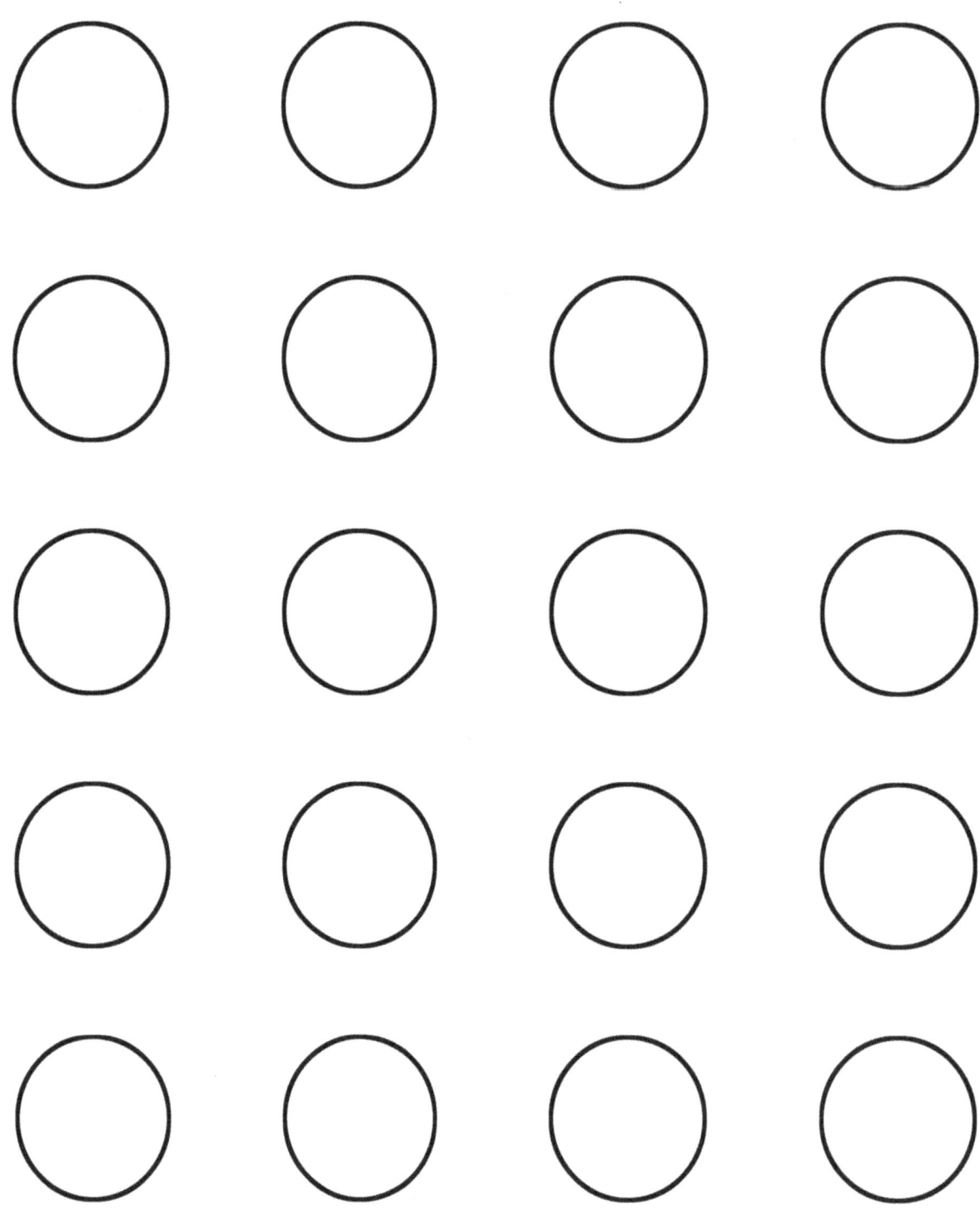

1. lime green, 2. dodger blue / corn flower blue, 3. aqua, 4. yellow, 5. red, 6. orange / golden rod, 7. Lime

$4-2$ $8-6$ $3-1$ $1+1$ $4-2$ $1+1$
$5-3$ $6-5$ $6-5$ $5-4$
$7-5$ $1+1$ $8-7$ $7-2$ $8-3$
$8-6$ $7-1$
$6-4$ $3-1$ $7-5$ $4+1$ $2+3$
$5+1$ $1+1$
$4-2$ $3-1$ $6-1$ $3+2$
$7-5$ $8-6$ $1+4$ $7-2$
$6-4$ $4-2$
$5-3$ $3-1$ $8-3$ $6-2$
$4-2$ $7-5$ $2+2$
$6-4$ $5-1$ $3+1$
$5-4$ $8-6$ $6-5$
$1+1$ $3-2$
$4-2$ $2-1$ $4+2$ $5+1$ $7-1$
$4-3$ $7-6$
$8-7$ $9-8$
$3-1$ $5-3$ $4-3$ $7-6$ $8-7$

$4-3$ $3-2$ $2-1$ $7-1$
$3-2$ $1+5$ $2+4$ $5+1$ $4+3$ $3+4$
$2+5$
$6-2$ $3+1$ $7-3$ $2+2$ $8-1$
$6-3$ $4-1$ $8-4$ $2+5$
$8-4$ $5-2$ $8-5$ $1+3$ $9-2$ $6+1$ $5+2$
$5-1$ $2+2$ $8-4$ $3+1$ $4+3$ $2+5$
$3+1$ $4-1$ $1+2$ $5-1$ $1+3$ $3+4$
$6-2$ $2+1$ $7-4$ $7-3$ $6+1$
$5-1$ $6-2$ $2+2$ $3+1$ $1+6$
$1+3$ $1+3$ $7-3$ $8-4$ $8-2$
$7-3$ $6-1$ $3+2$ $8-3$ $5+1$
$8-4$ $2+3$ $8-5$ $1+4$
$2+2$ $7-2$ $6-1$ $3+4$
$5-1$ $6-2$ $3+1$
$3-2$ $1+5$ $4+2$
$3+3$ $1+6$ $6+1$
$2+4$ $8-1$
$8-2$ $5+2$

Test Your Color

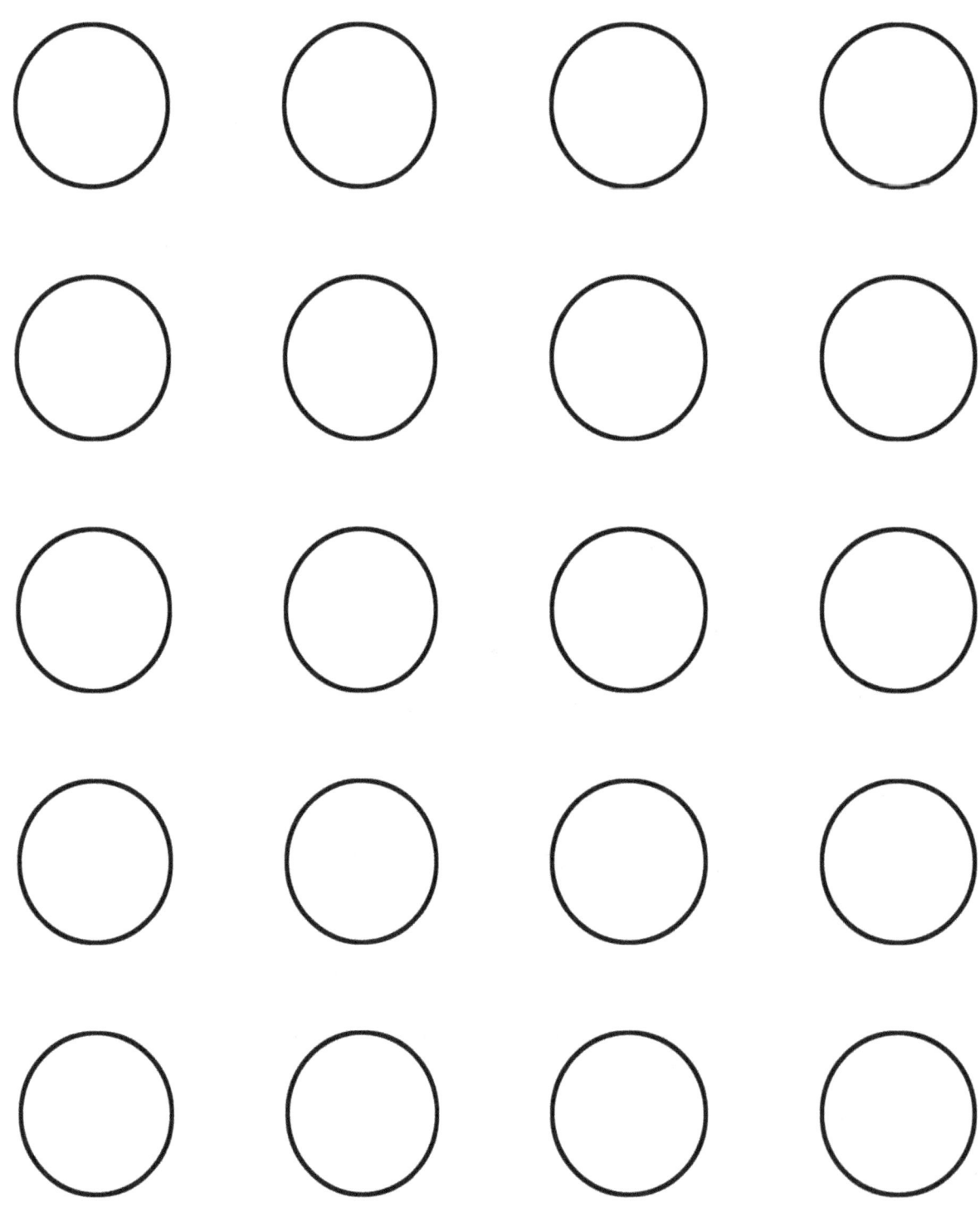

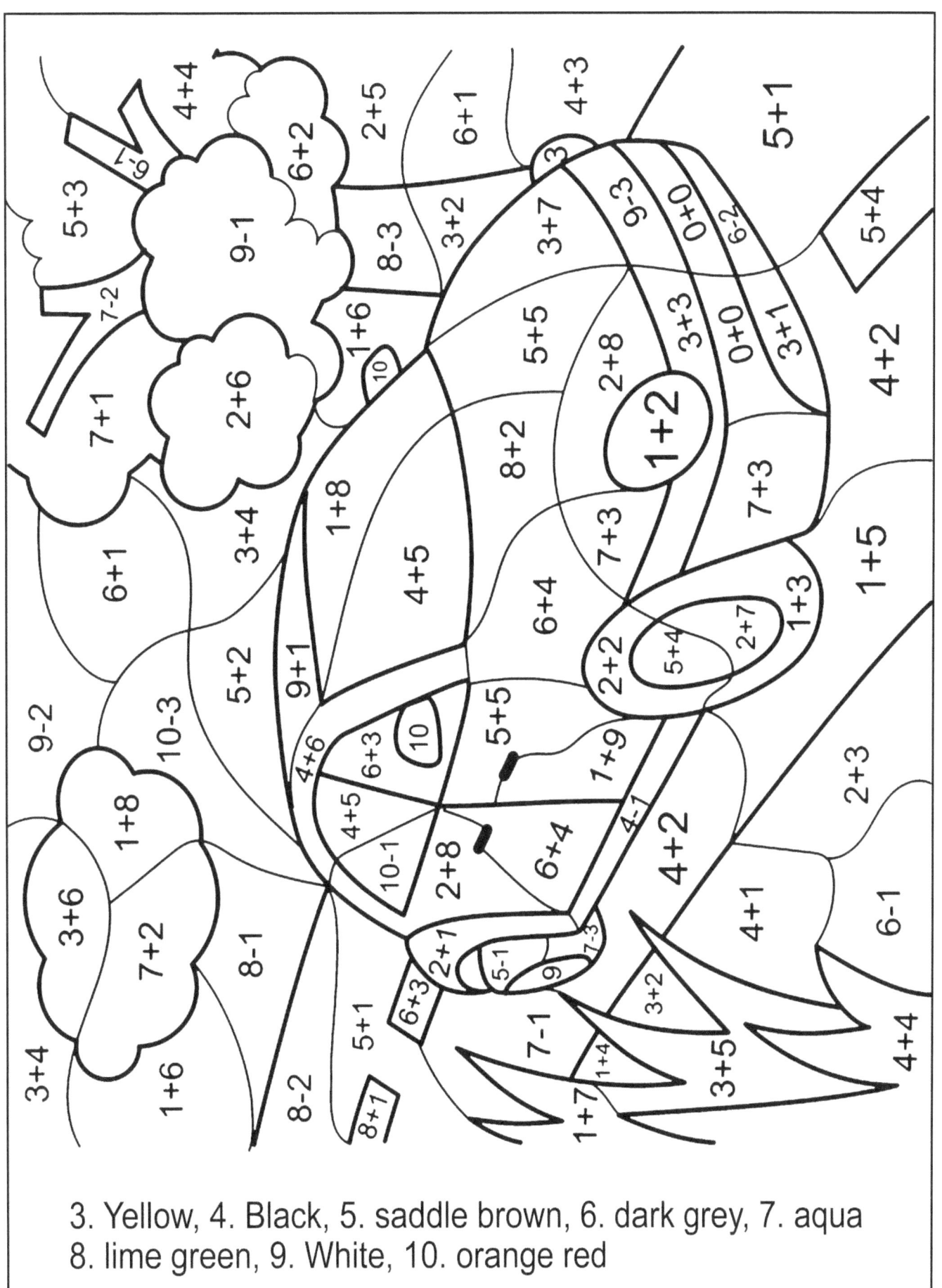

3. Yellow, 4. Black, 5. saddle brown, 6. dark grey, 7. aqua
8. lime green, 9. White, 10. orange red

Test Your Color

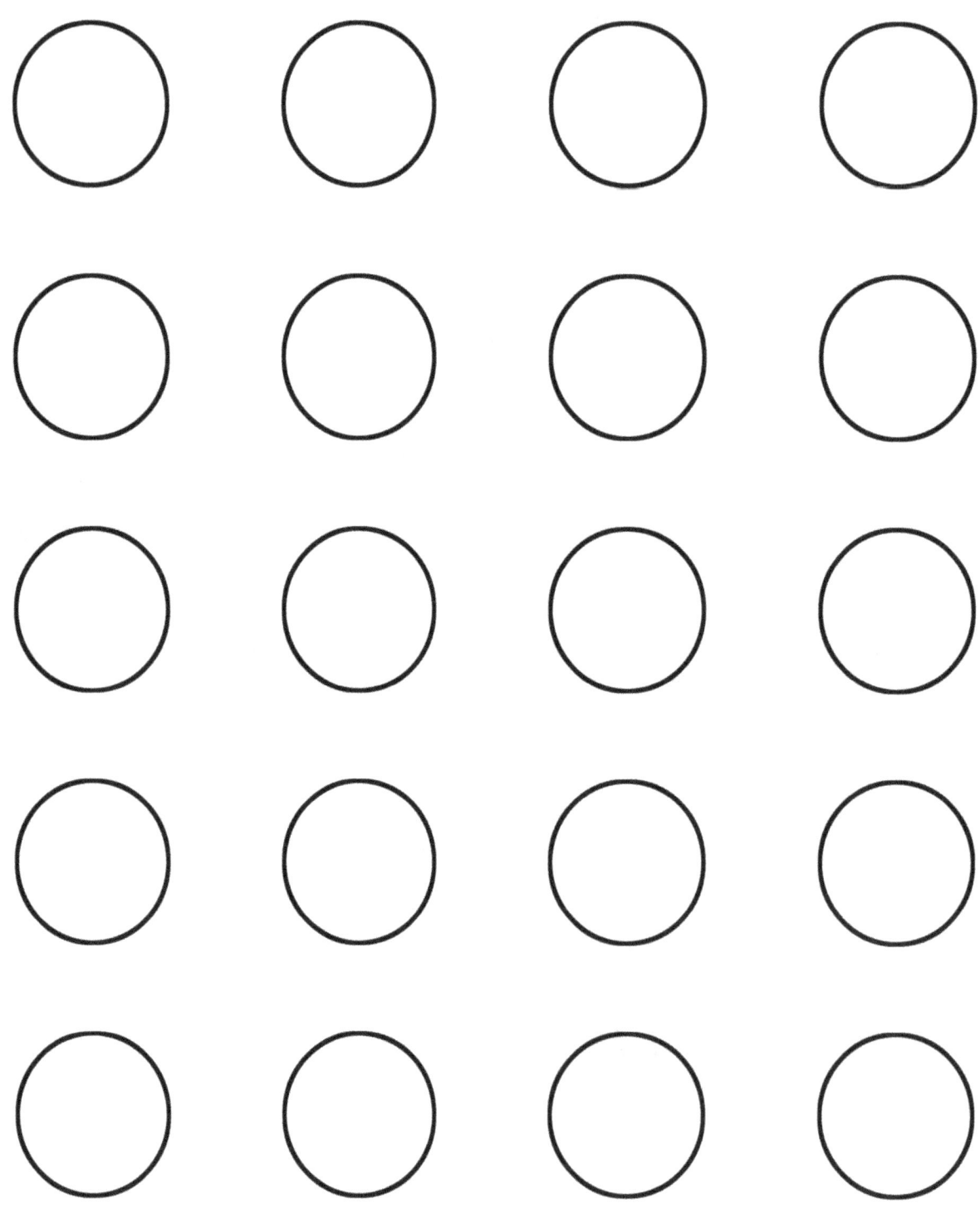

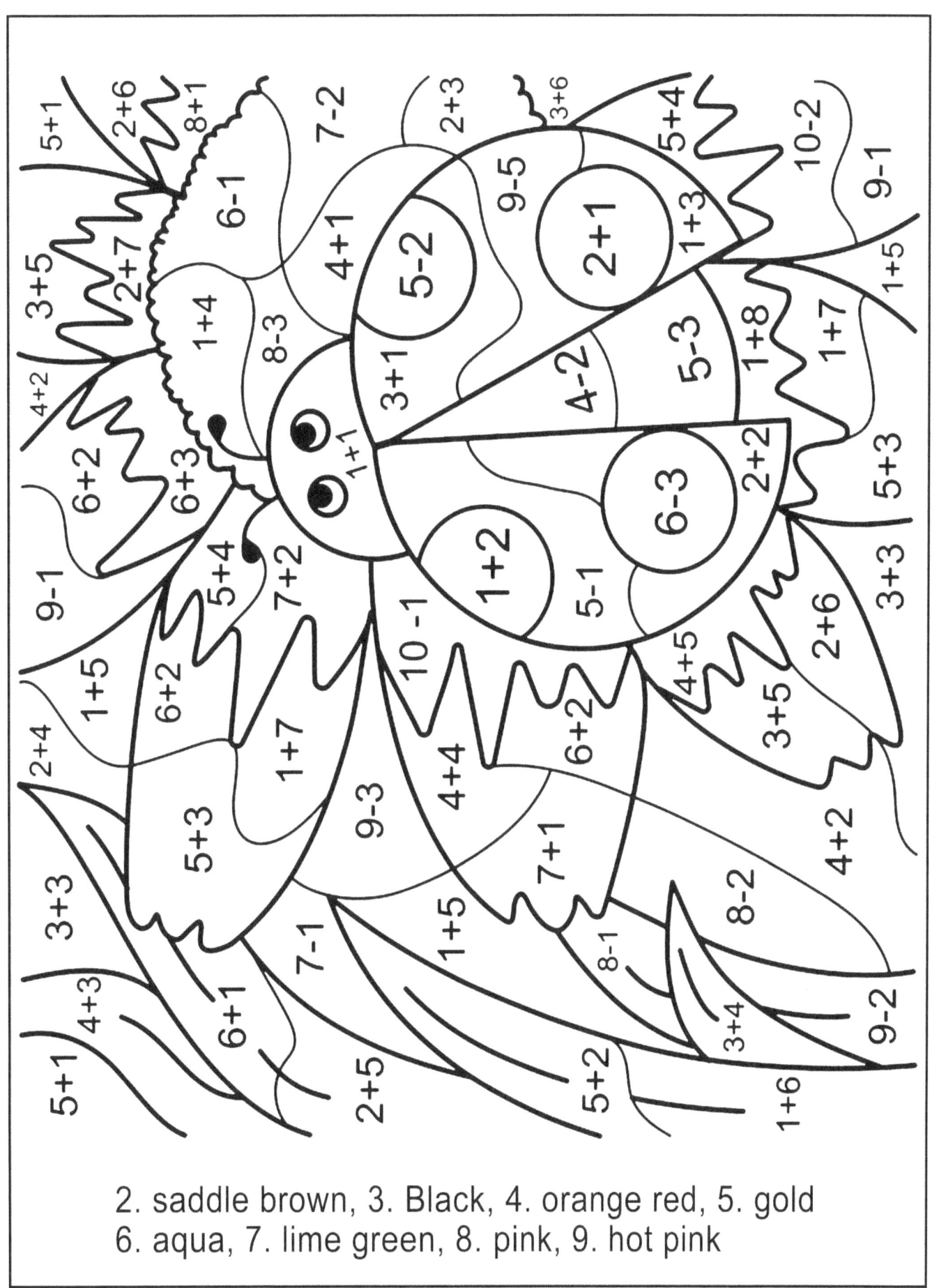

2. saddle brown, 3. Black, 4. orange red, 5. gold
6. aqua, 7. lime green, 8. pink, 9. hot pink

Test Your Color

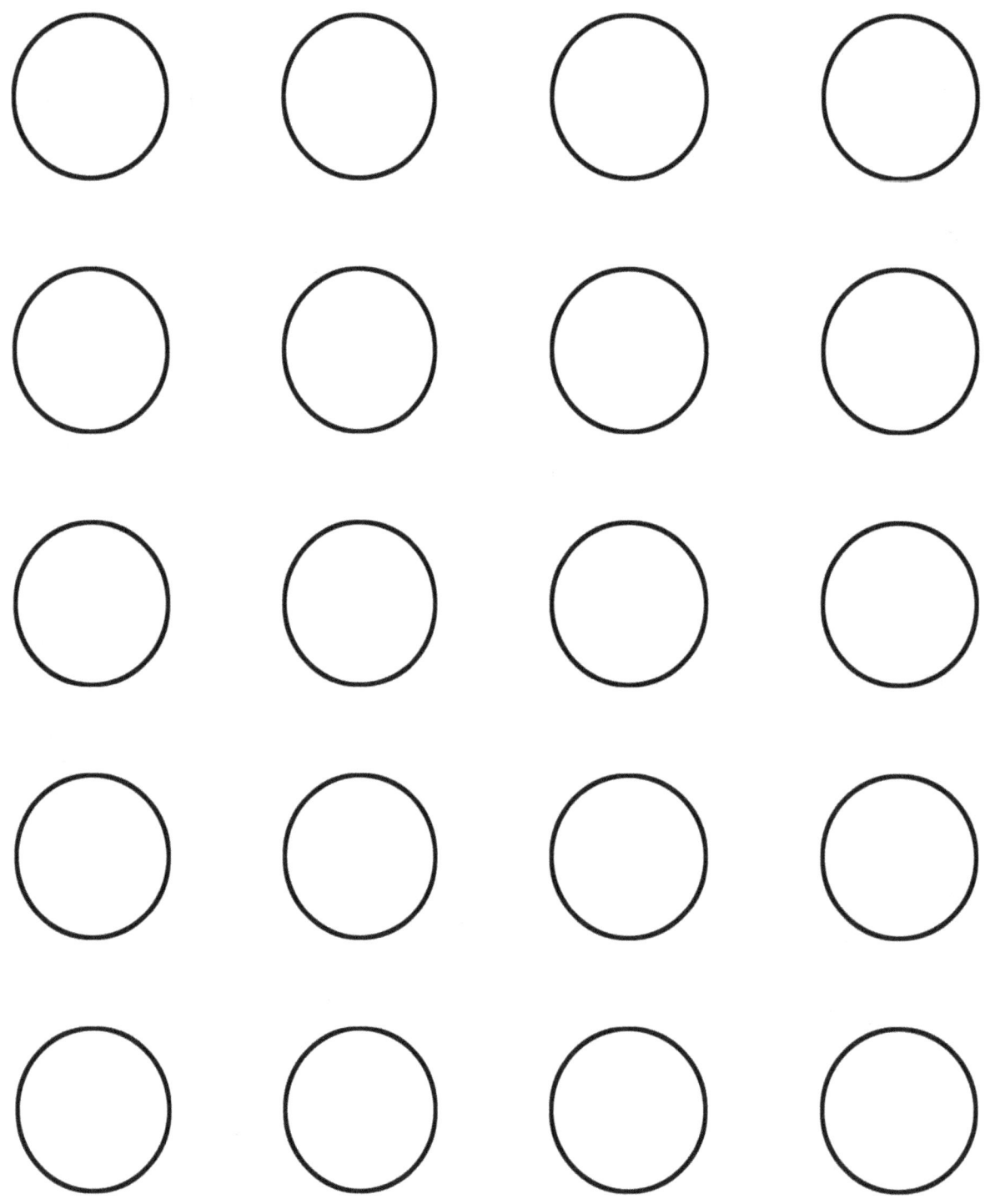

5-3= red, 2+3= peru, 3+6= aqua, 8-4= yellow
7+1= dark red, 3-0= sandy brown
4+3= dodger blue, 9-3= lawn green

Test Your Color

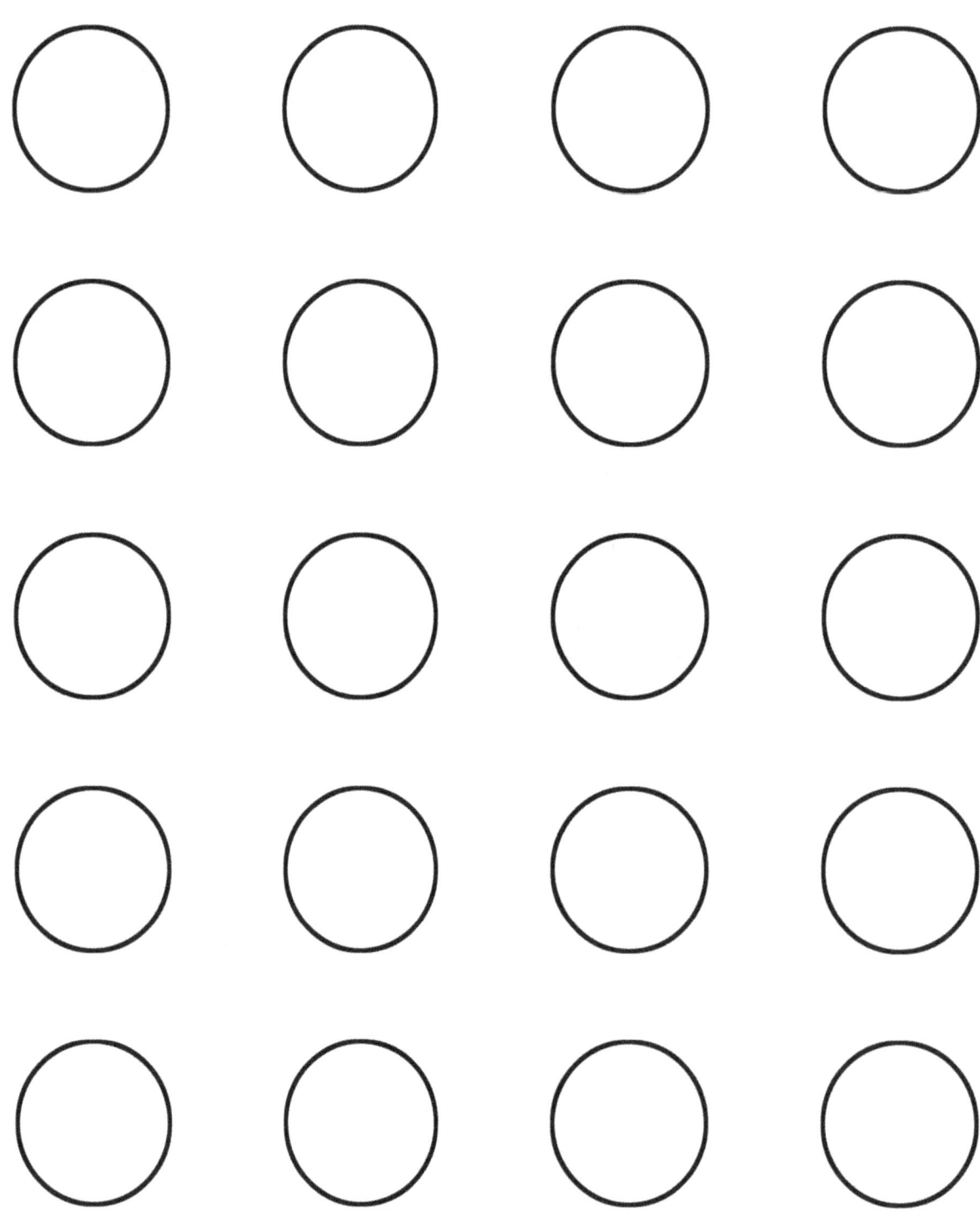

Test Your Color

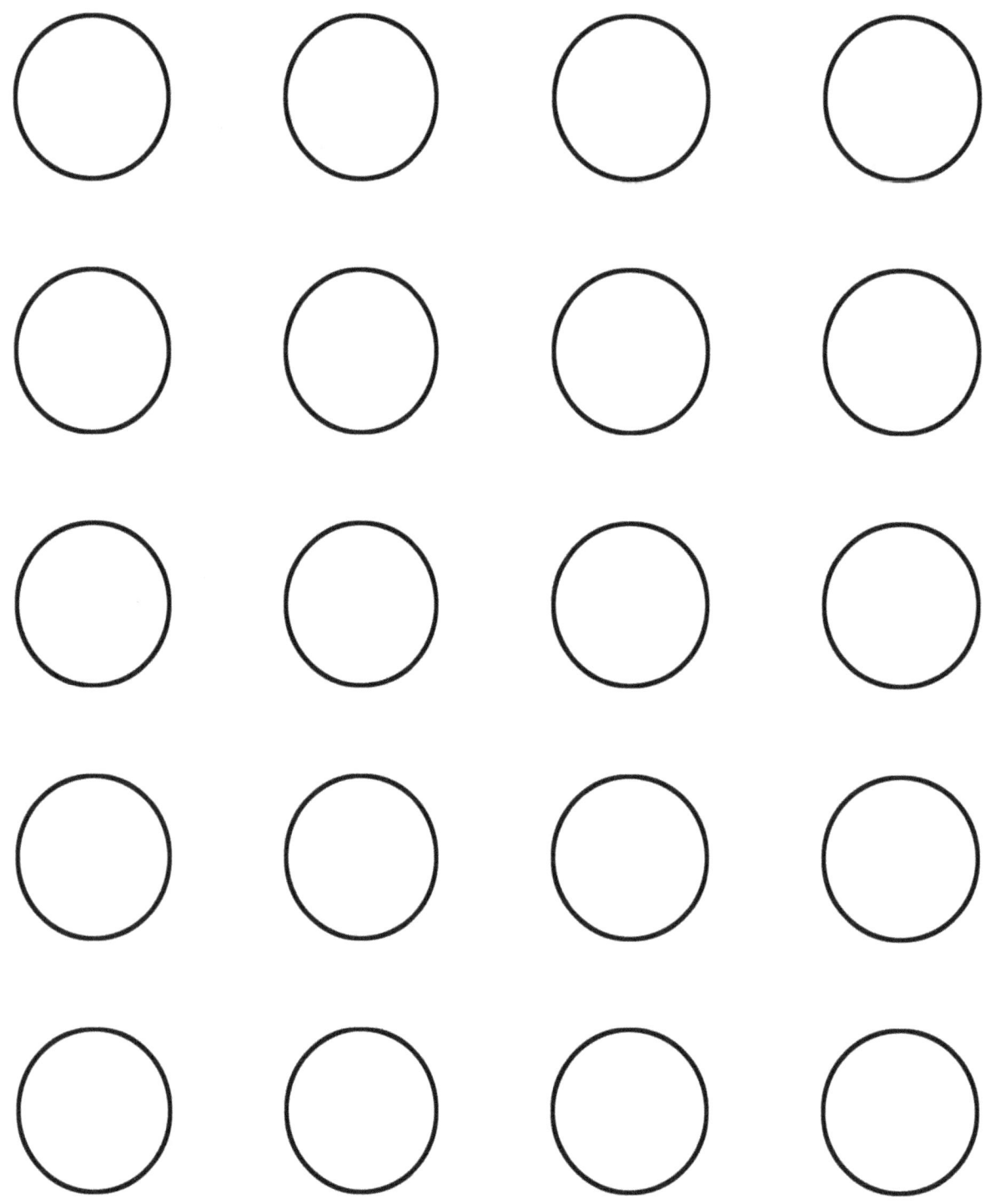

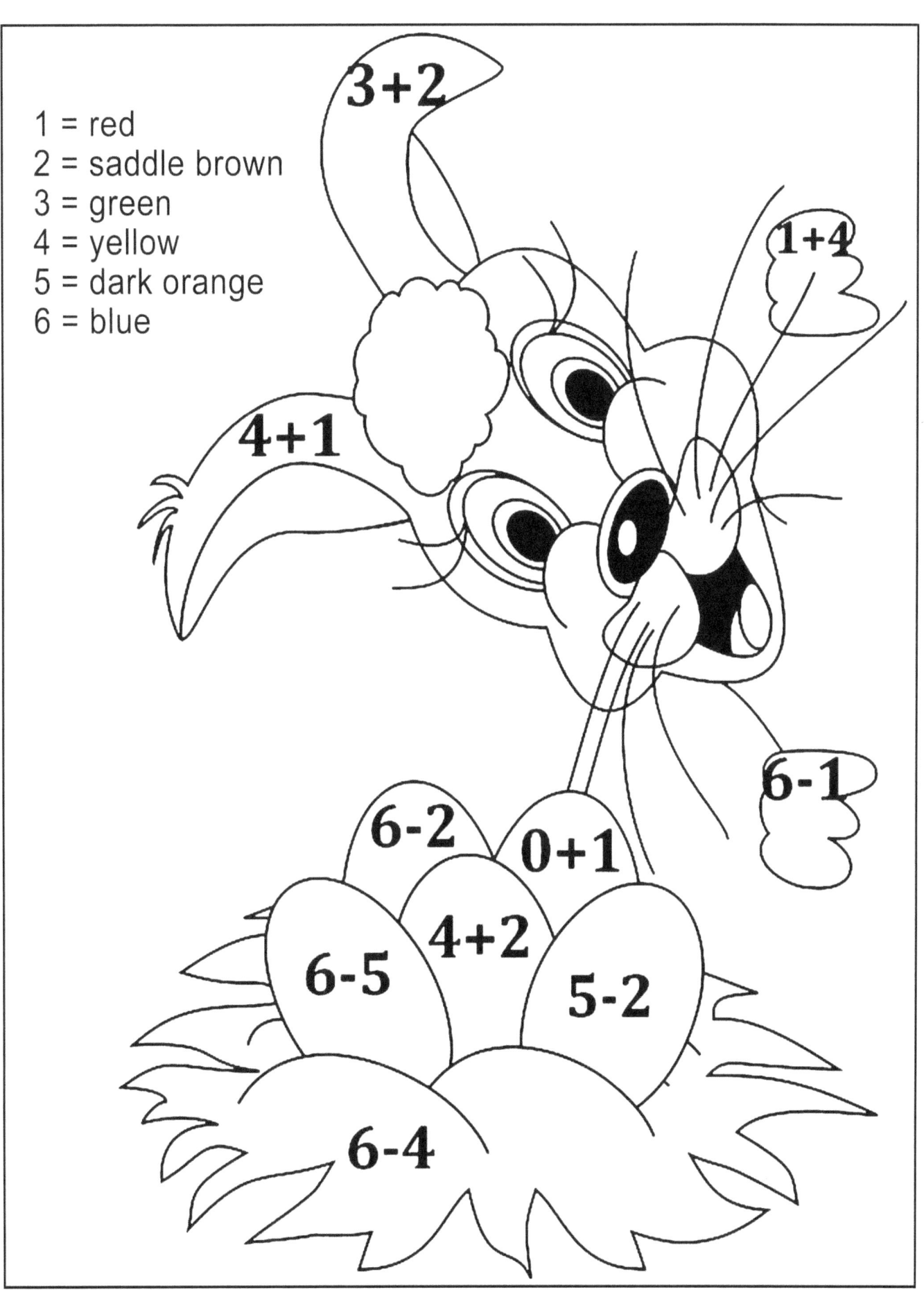

Test Your Color

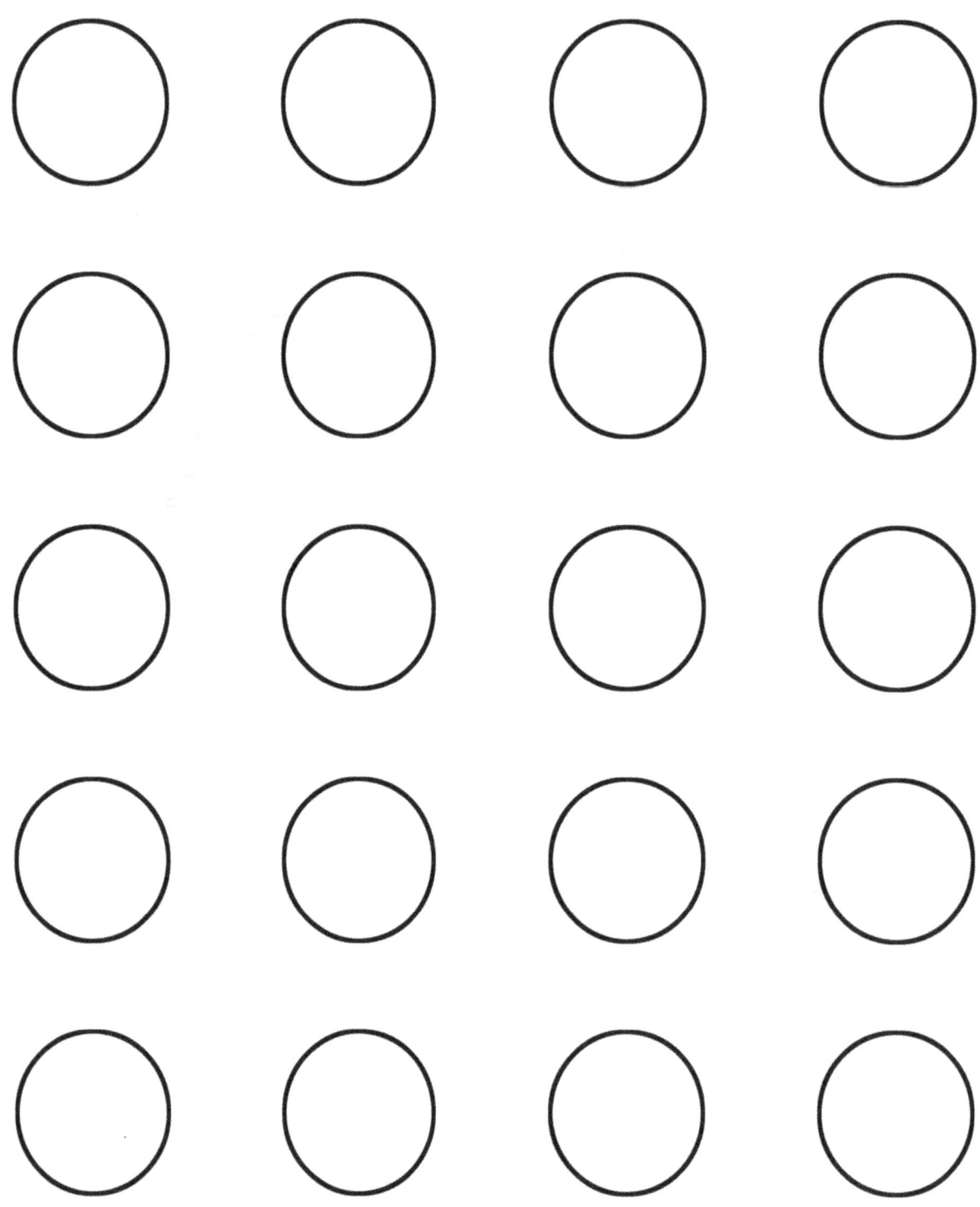

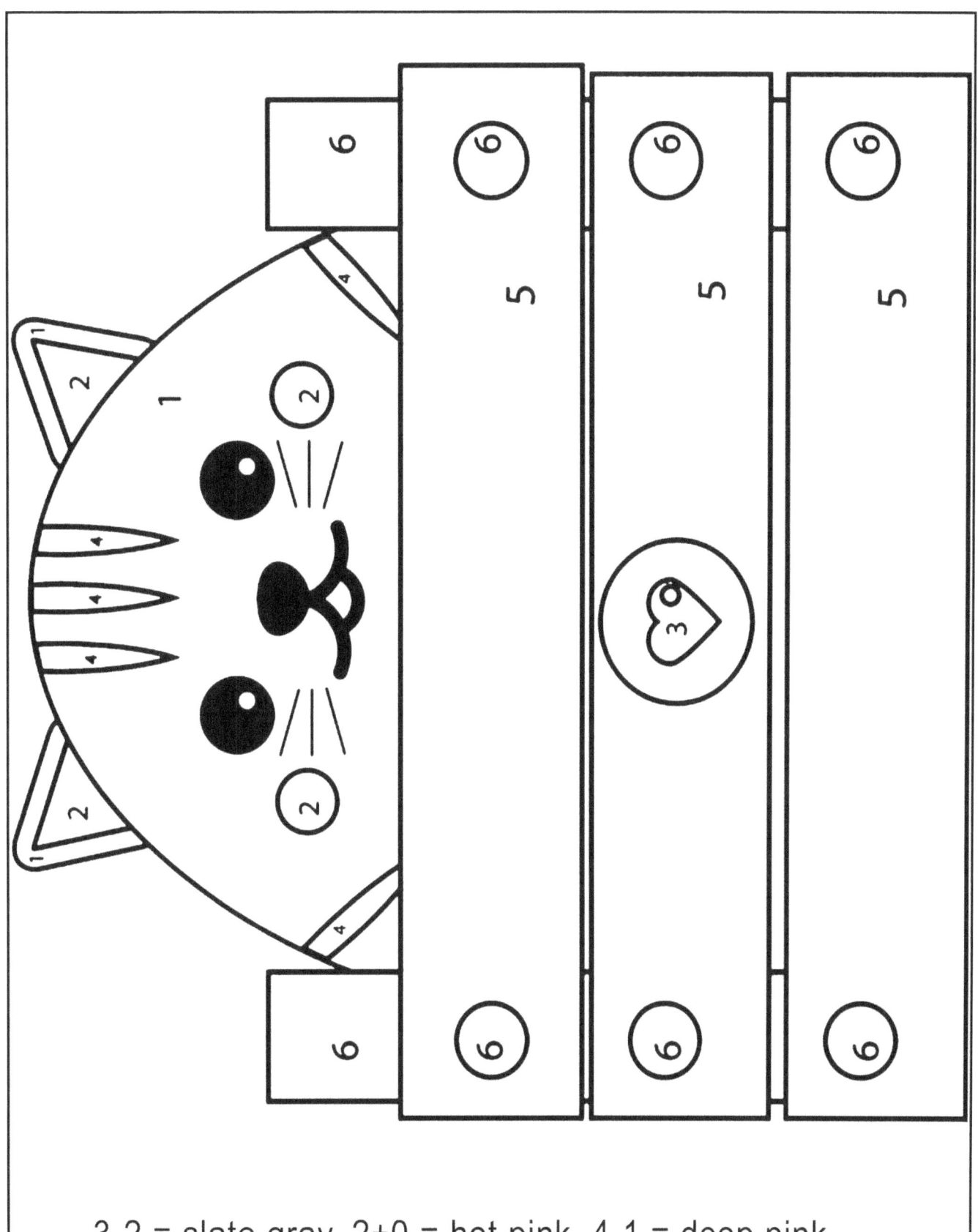

3-2 = slate gray, 2+0 = hot pink, 4-1 = deep pink
6-2 = black, 1+4 = medium purple, 9-3 = dark violet

Test Your Color

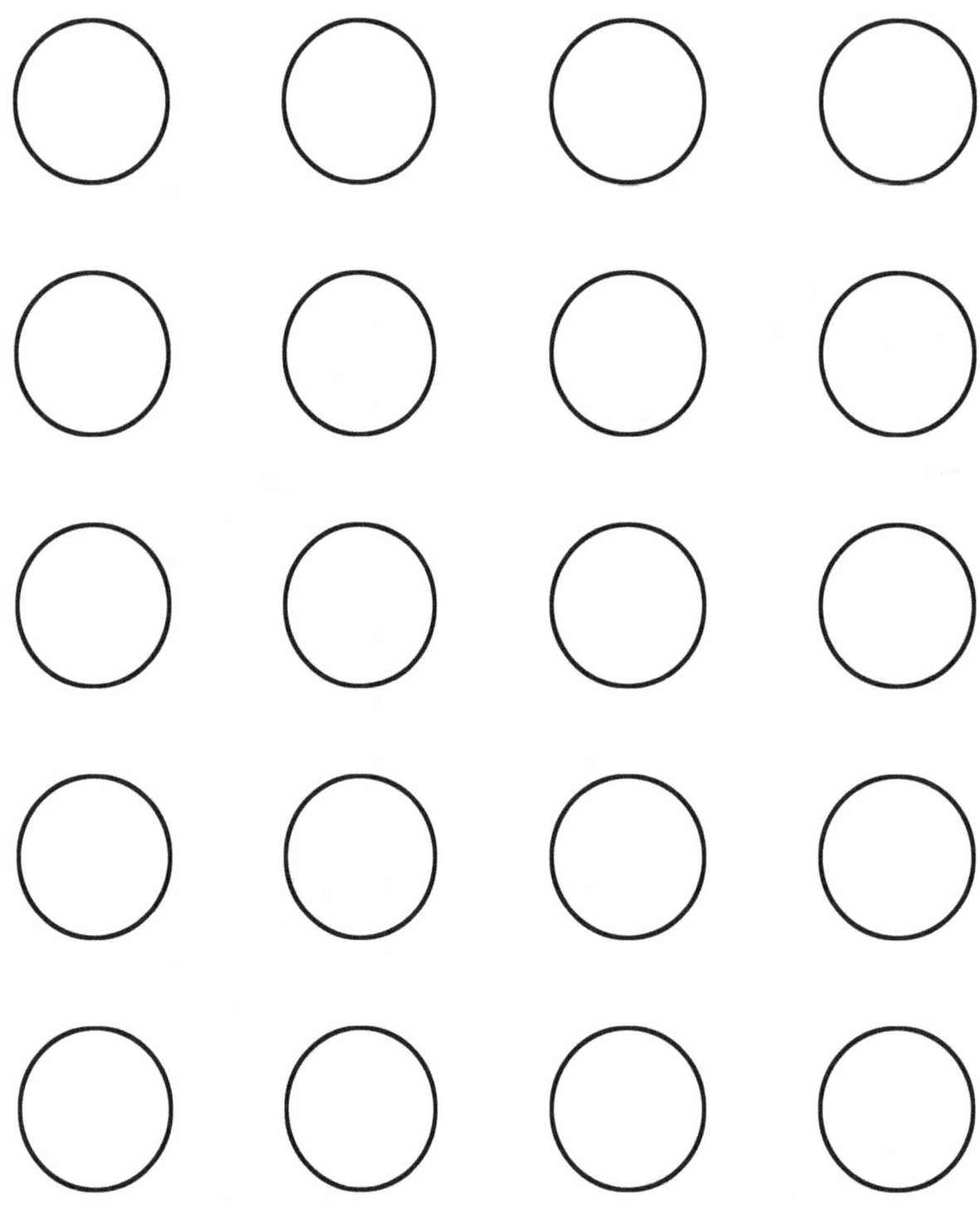

Color by numbers.

10-9 = dark orange, 2+2 = orange red, 3+4 = Purple
5-3 = Red, 10-5 = forest green, 2+6 = aqua, 2+1 = teal
7-1 = saddle brown, 1+8 = dark golden rod

Test Your Color

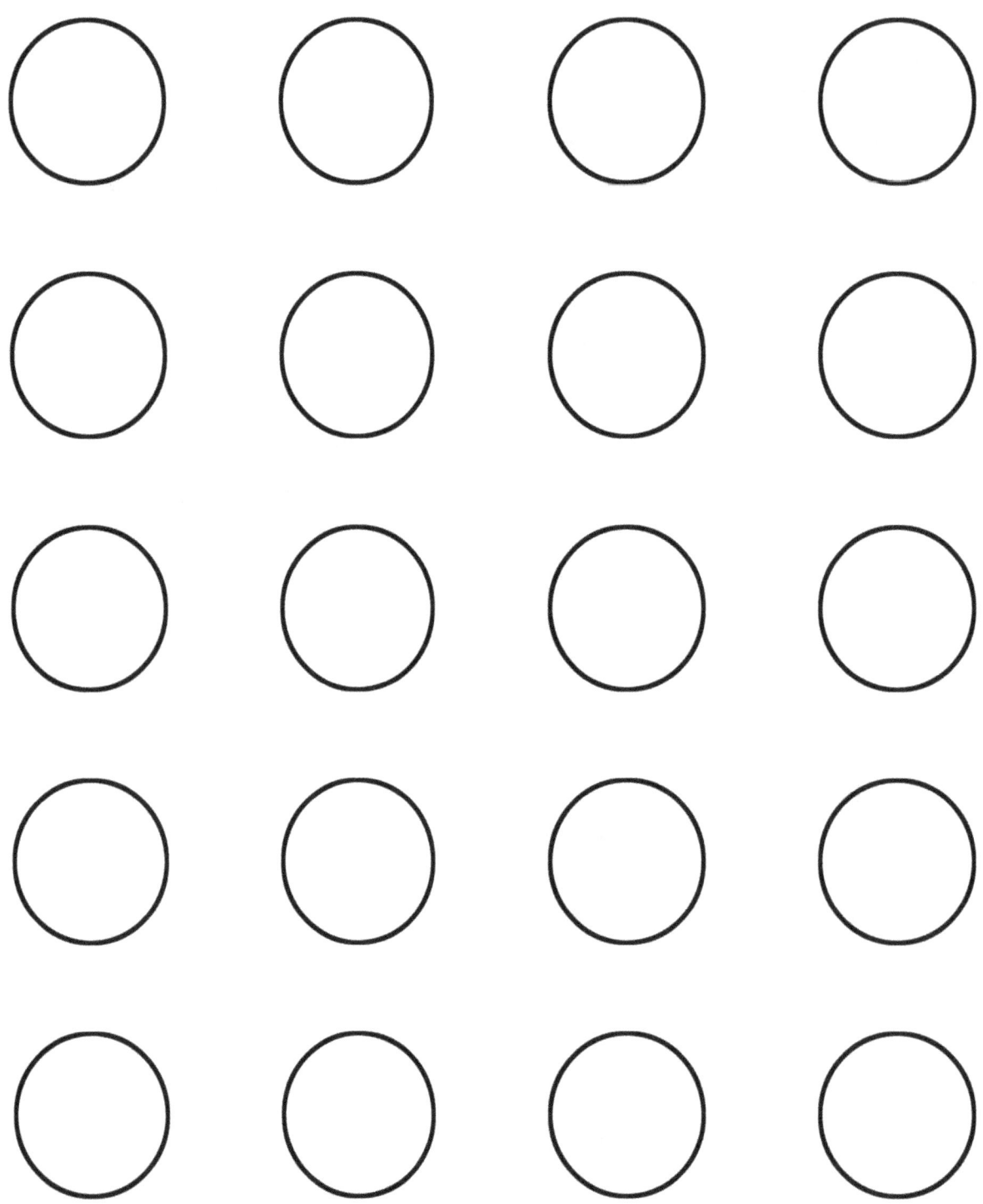

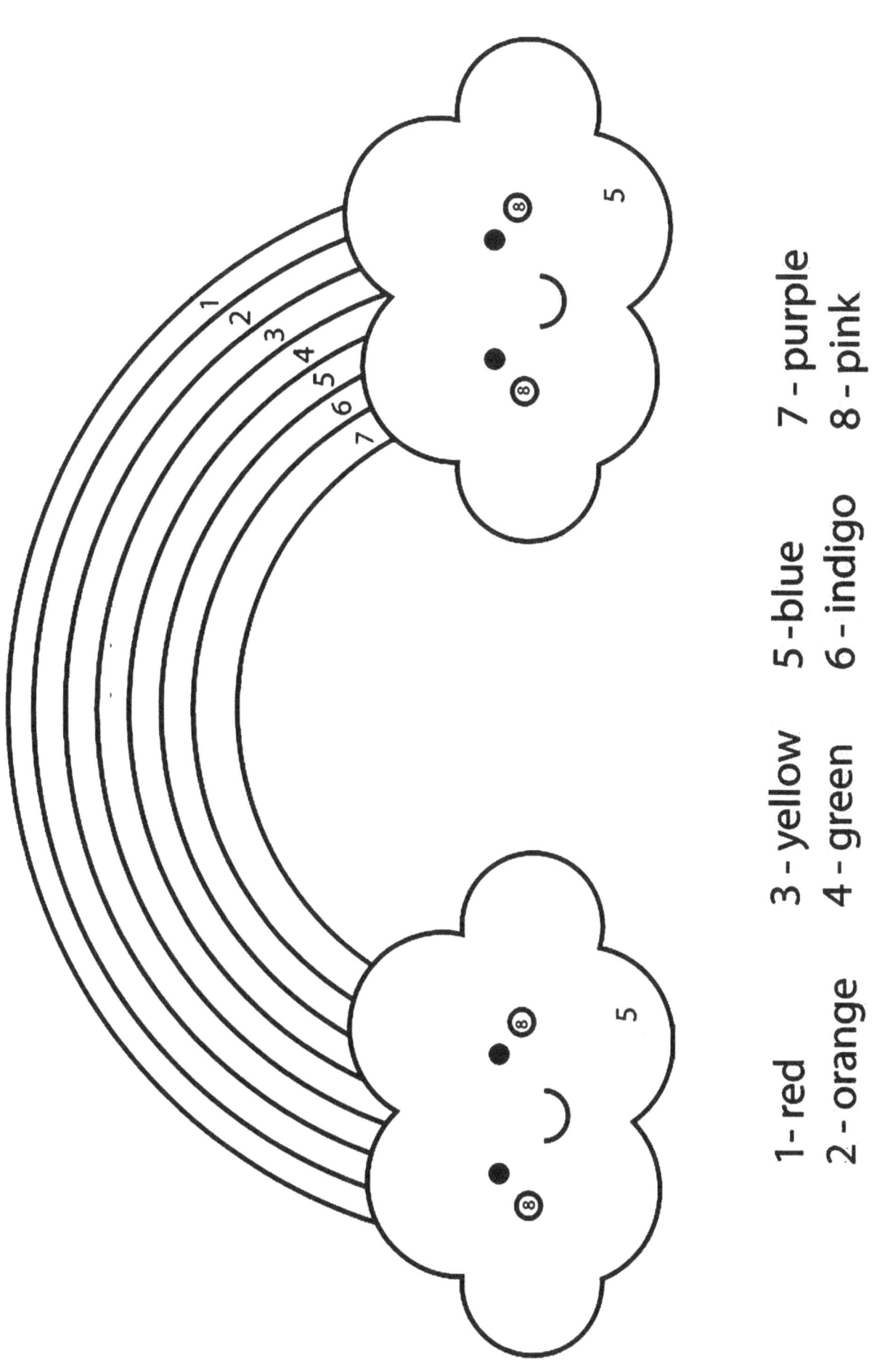

1 - red
2 - orange
3 - yellow
4 - green
5 - blue
6 - indigo
7 - purple
8 - pink

Test Your Color

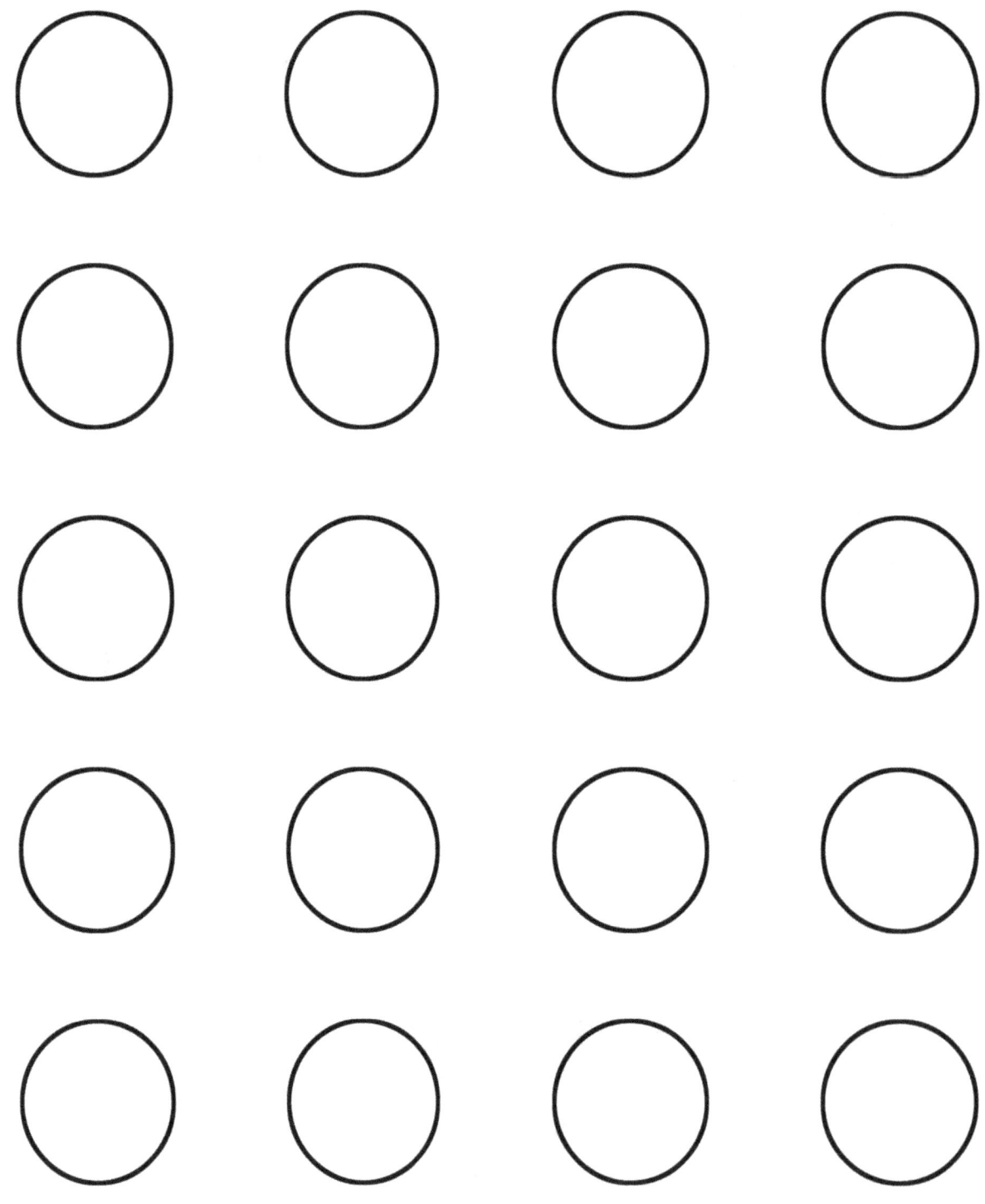

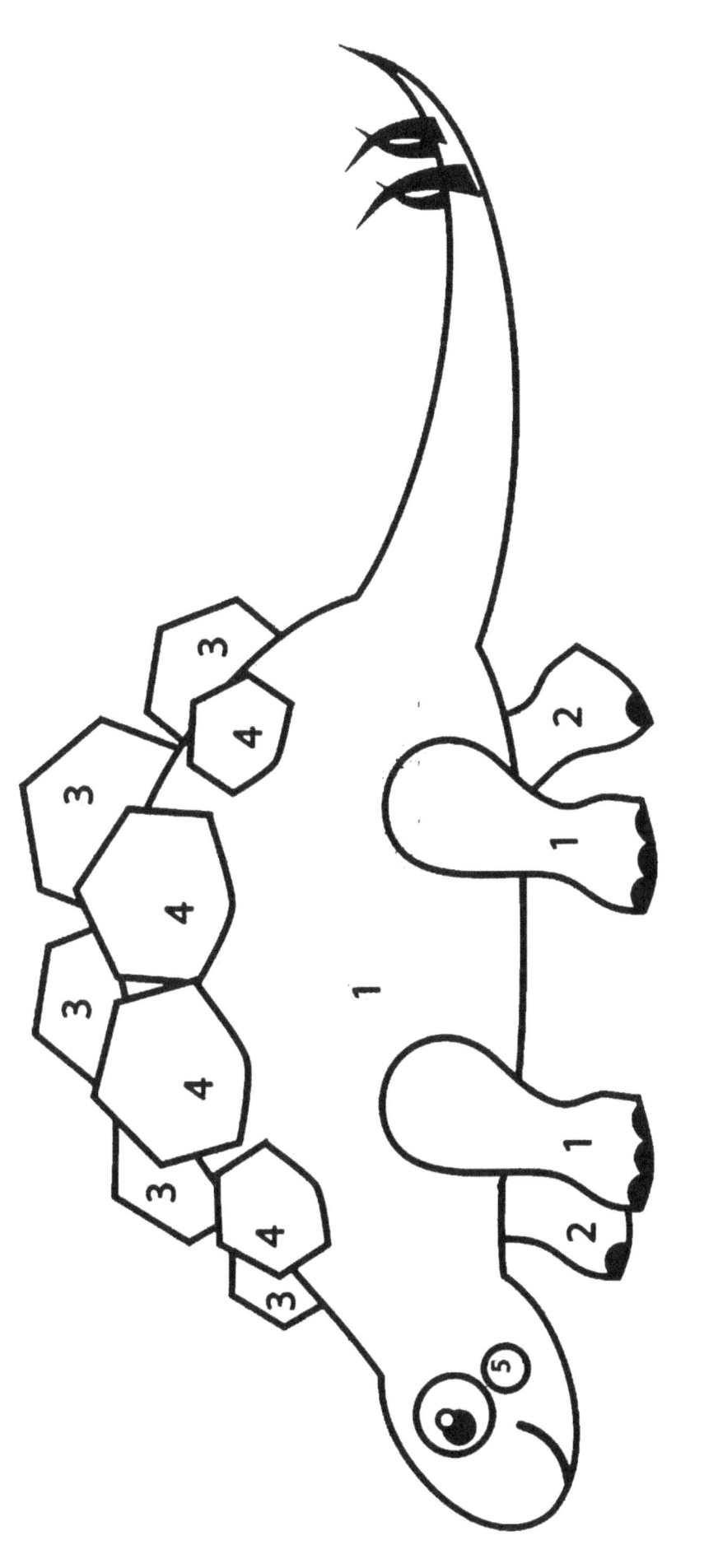

1 green
2 dark green

3 dark brown
4 brown
5 pink

Test Your Color

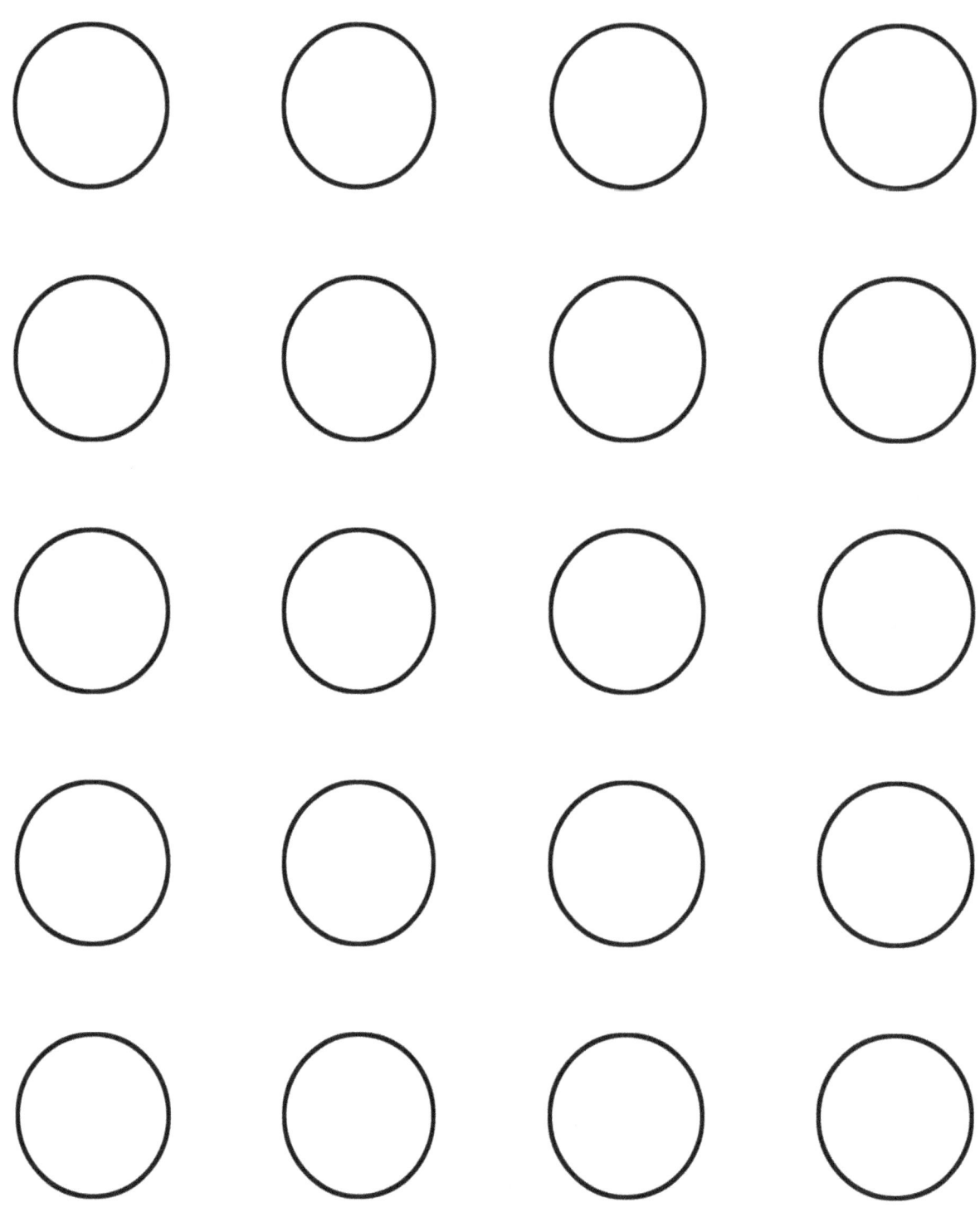

1 - light blue   2 - blue   3 - green   4 - dark green
5 - yellow   6 - orange   7 - brown   8 - pink

Test Your Color

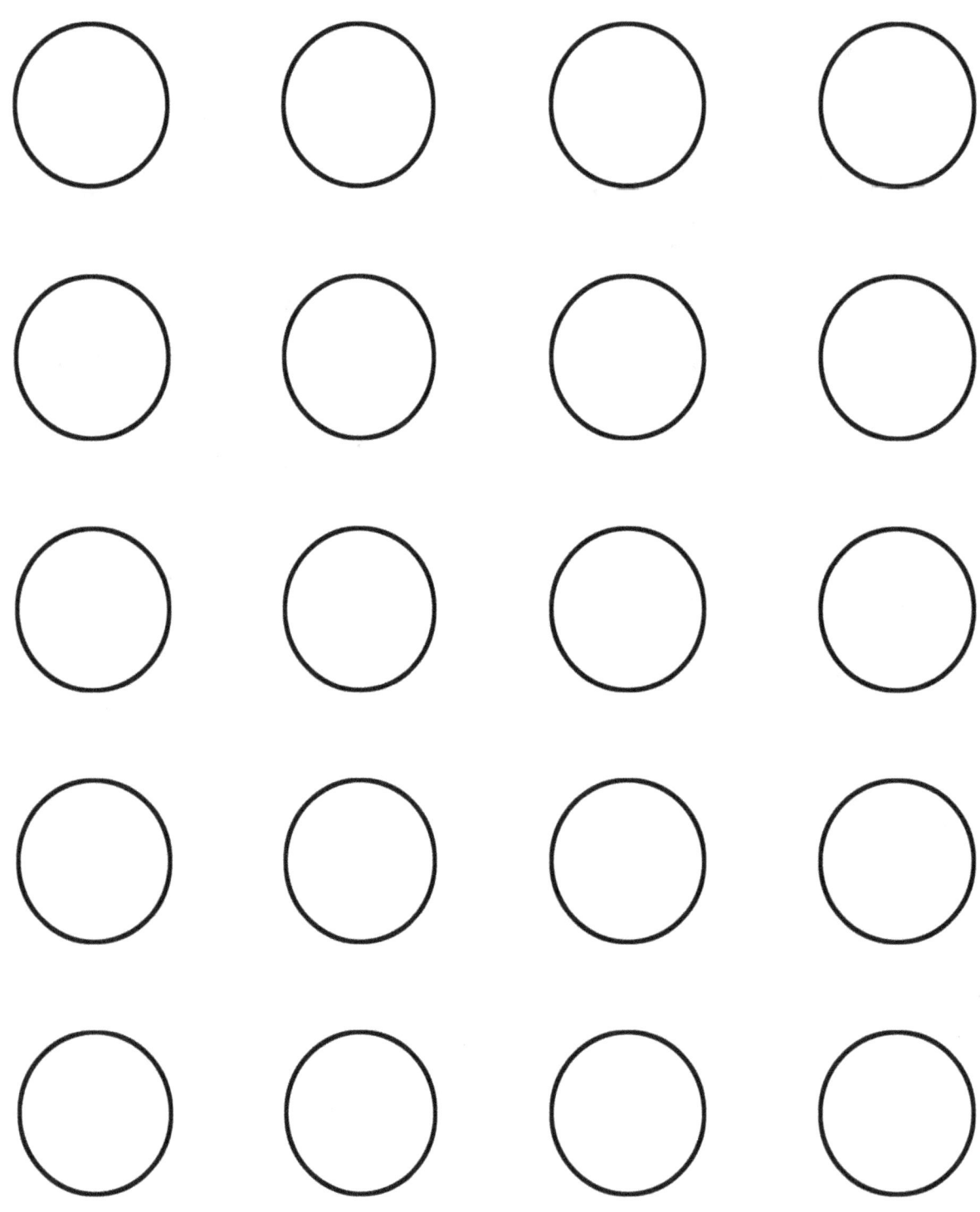

1 - light blue   2 - blue   3 - green   4 - dark green
5 - yellow   6 - orange   7 - gray   8 - brown

Test Your Color

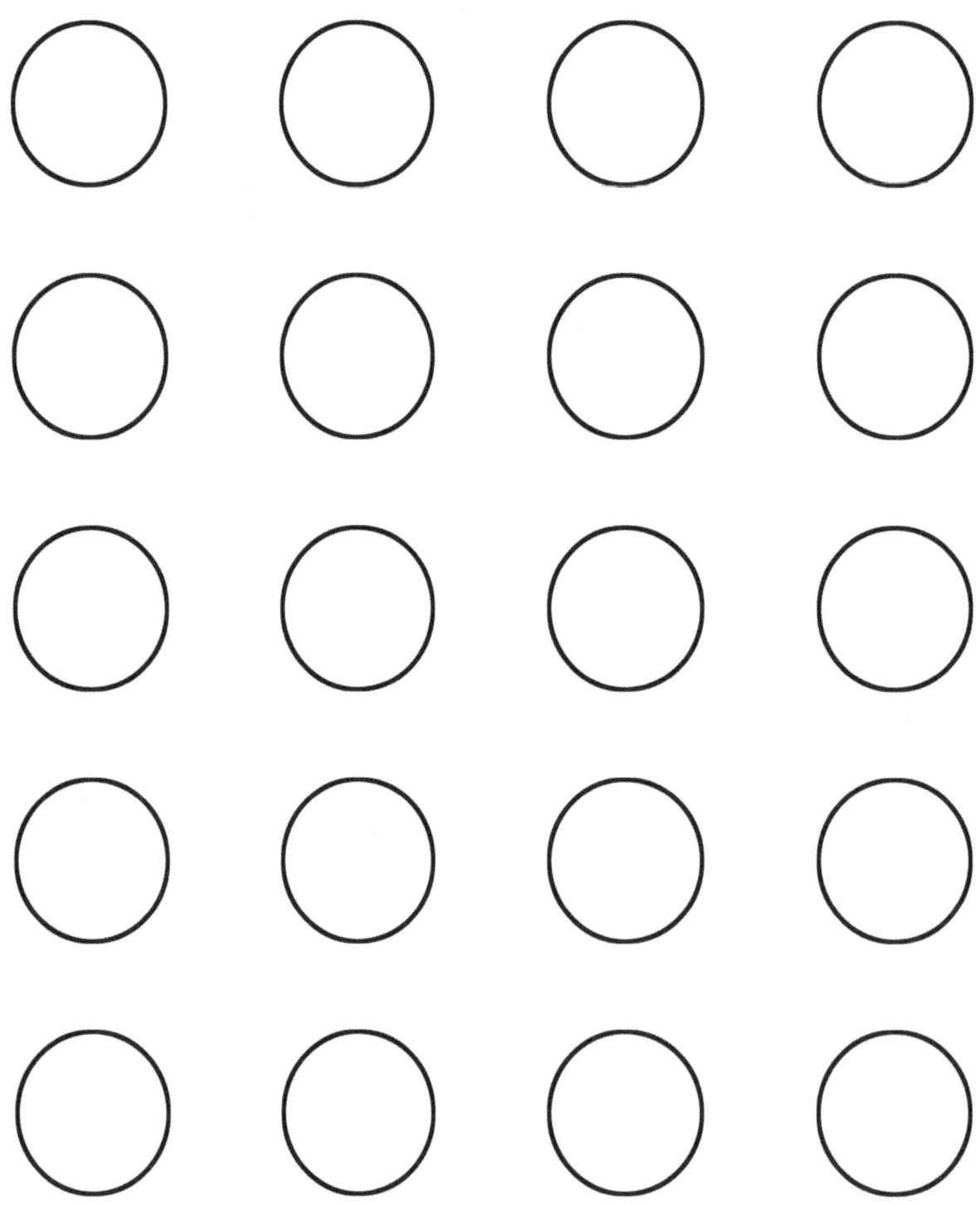

1 - light blue   2 - blue   3 - green   4 - dark green
5 - yellow   6 - orange   7 - brown

Test Your Color

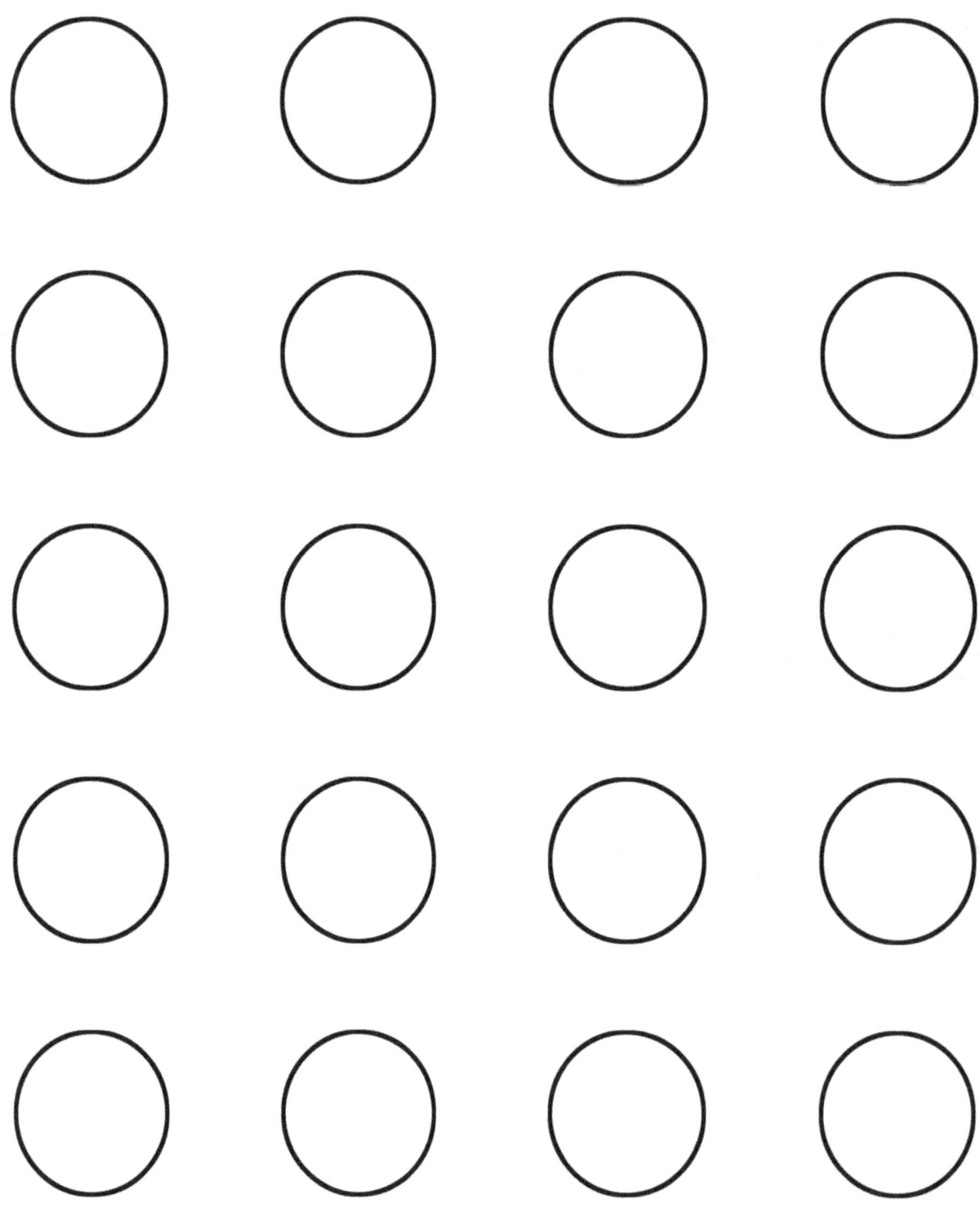

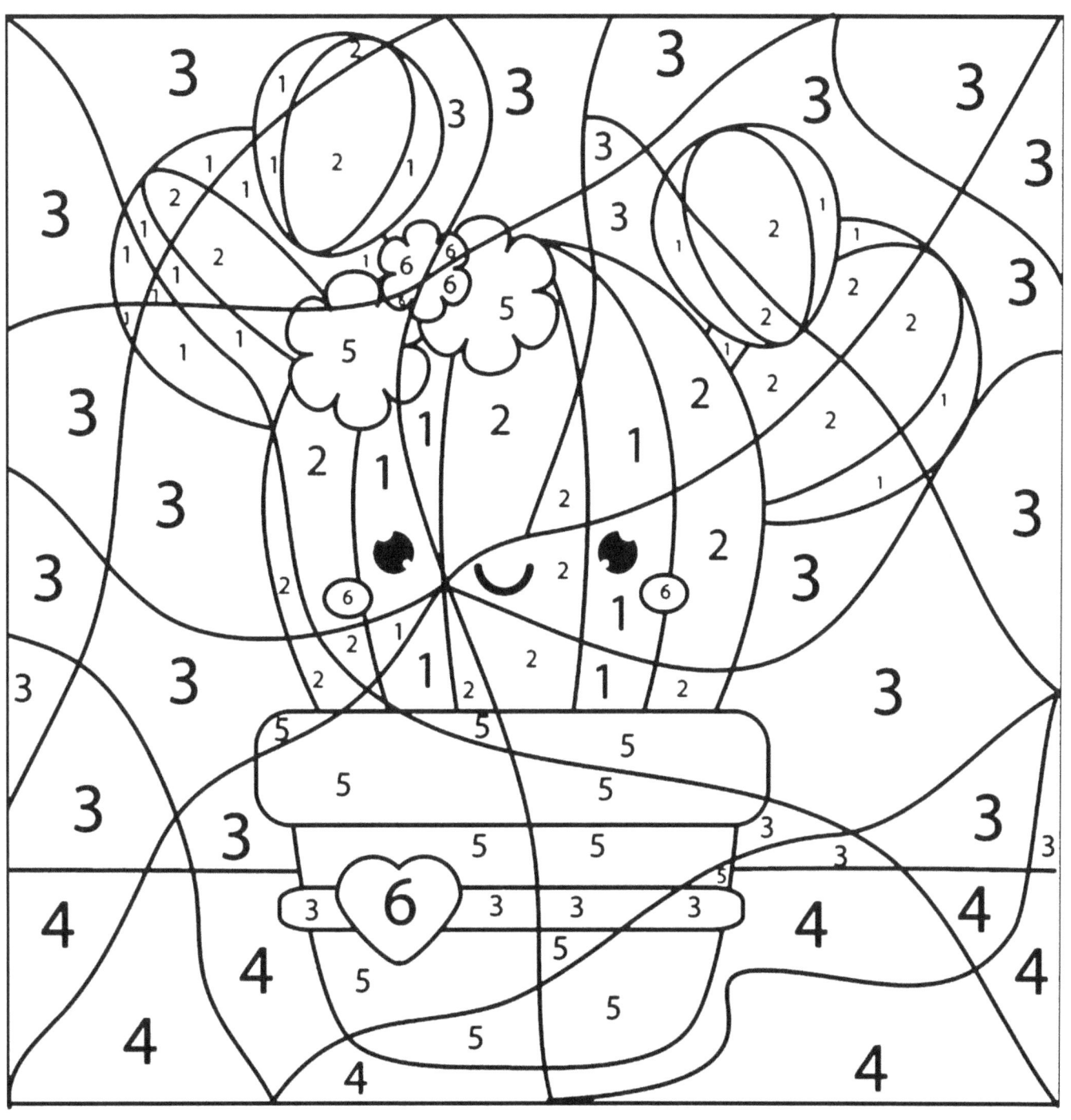

1 light green   3 blue   5 yellow
2 dark green   4 brown   6 orange

Test Your Color

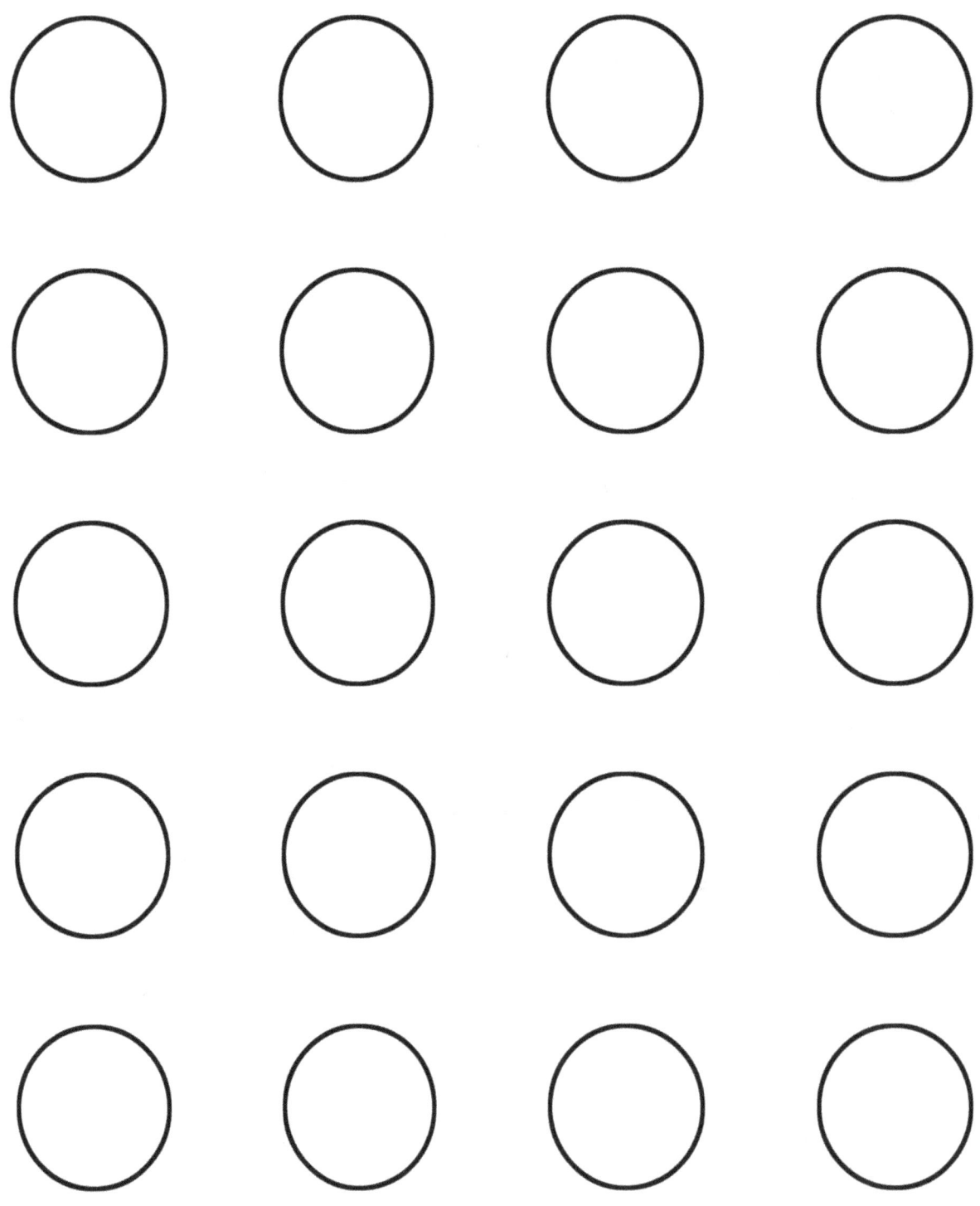

1 - light blue   2 - blue   3 - green   4 - dark green
5 - yellow   6 - orange   7 - brown

Test Your Color

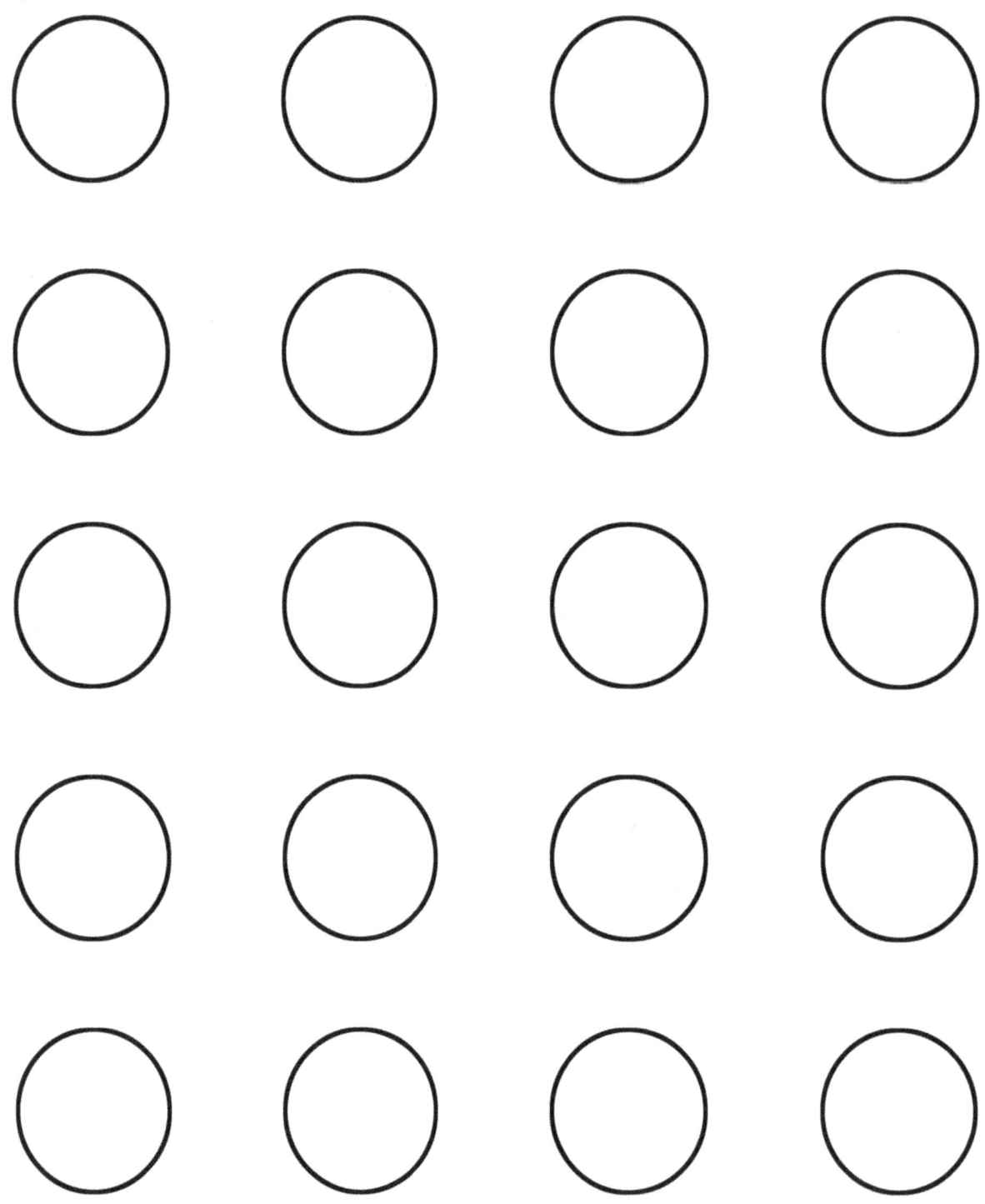

1 - yellow   2 - brown   3 - green   4 - dark green

Test Your Color

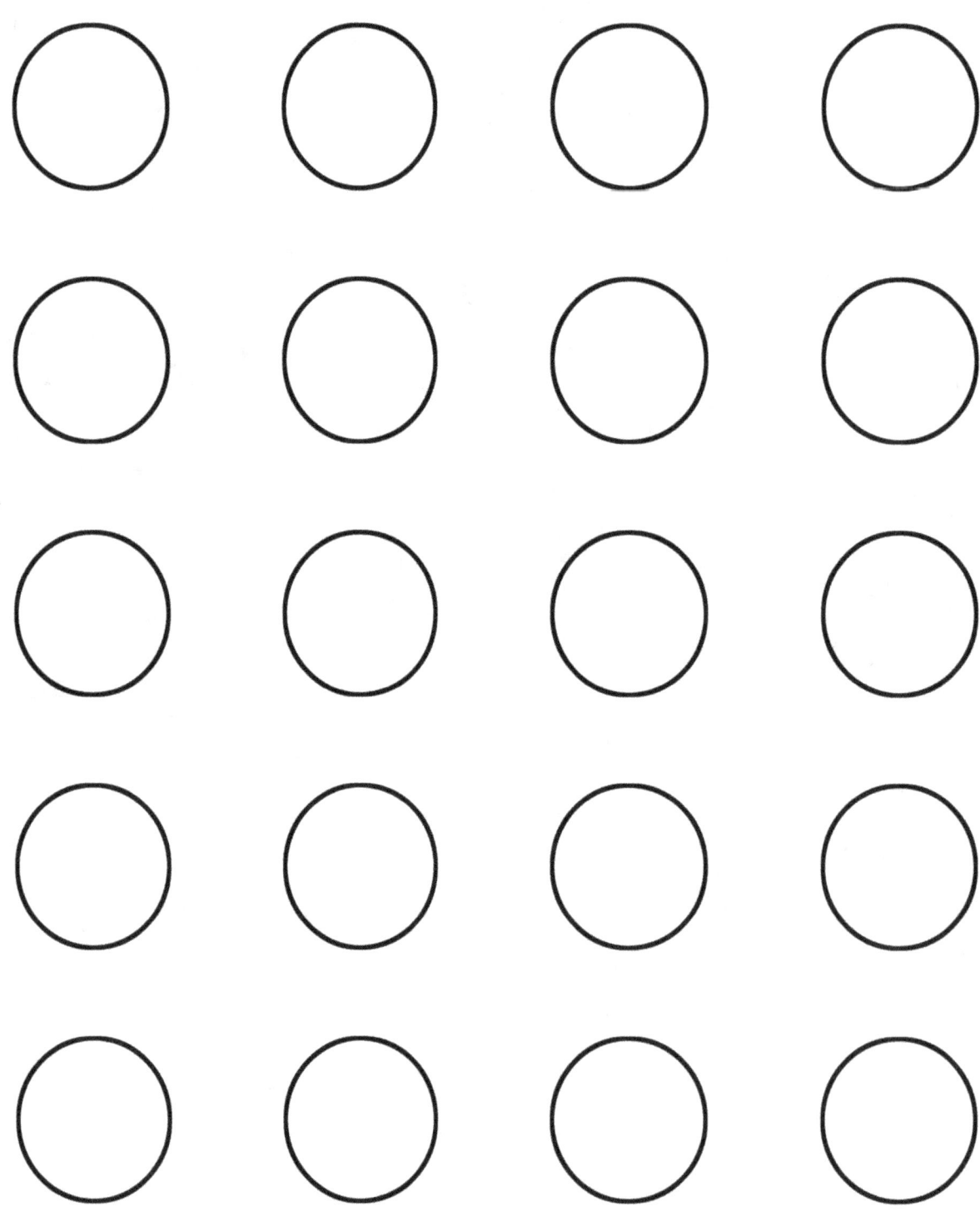

1 - light blue   2 - blue   3 - green   4 - dark green
5 - yellow   6 - orange   7 - gray   8 - brown   9 - black

Test Your Color

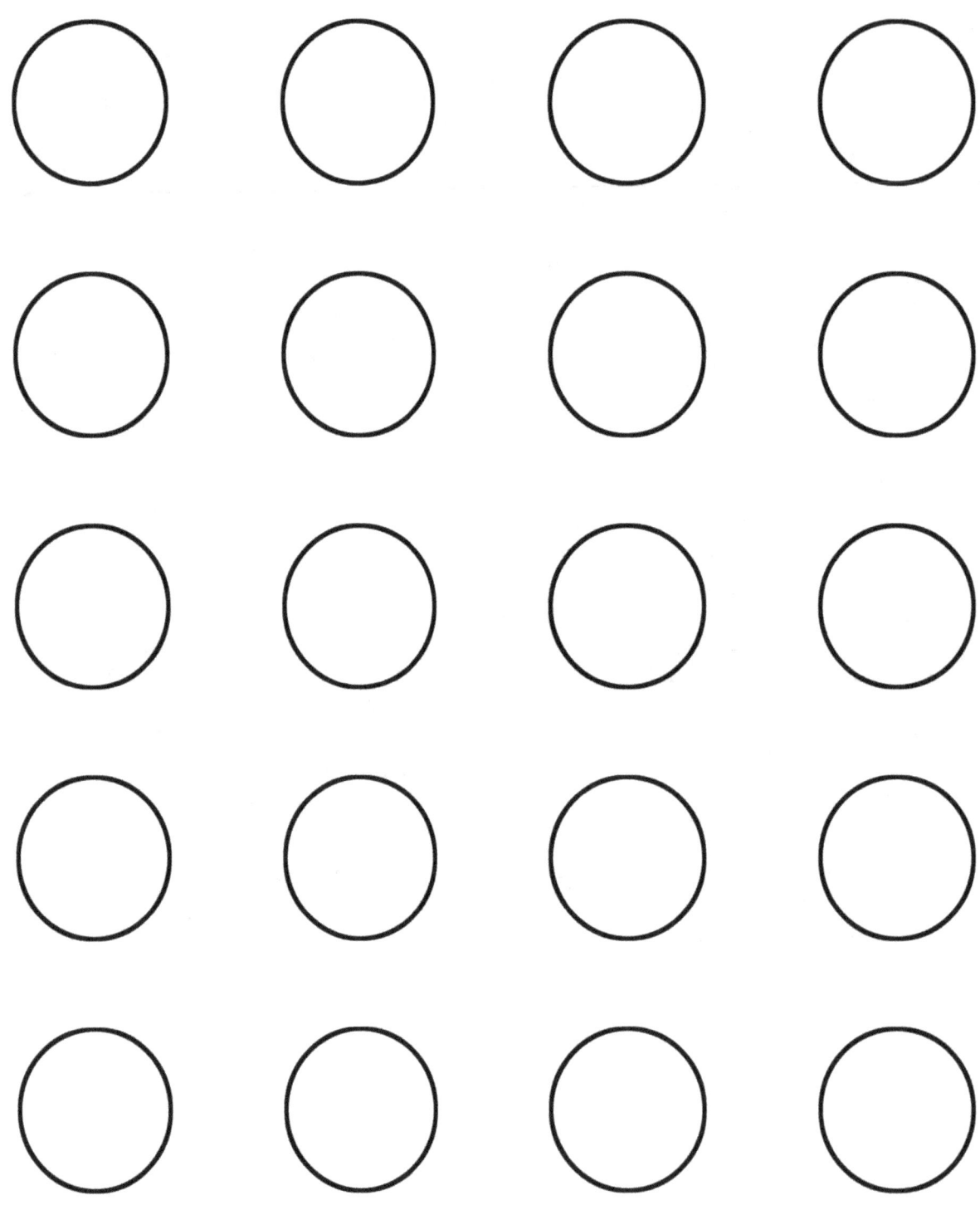

1 - light blue   2 - blue   3 - green
4 - yellow   5 - orange   6 - brown

Test Your Color

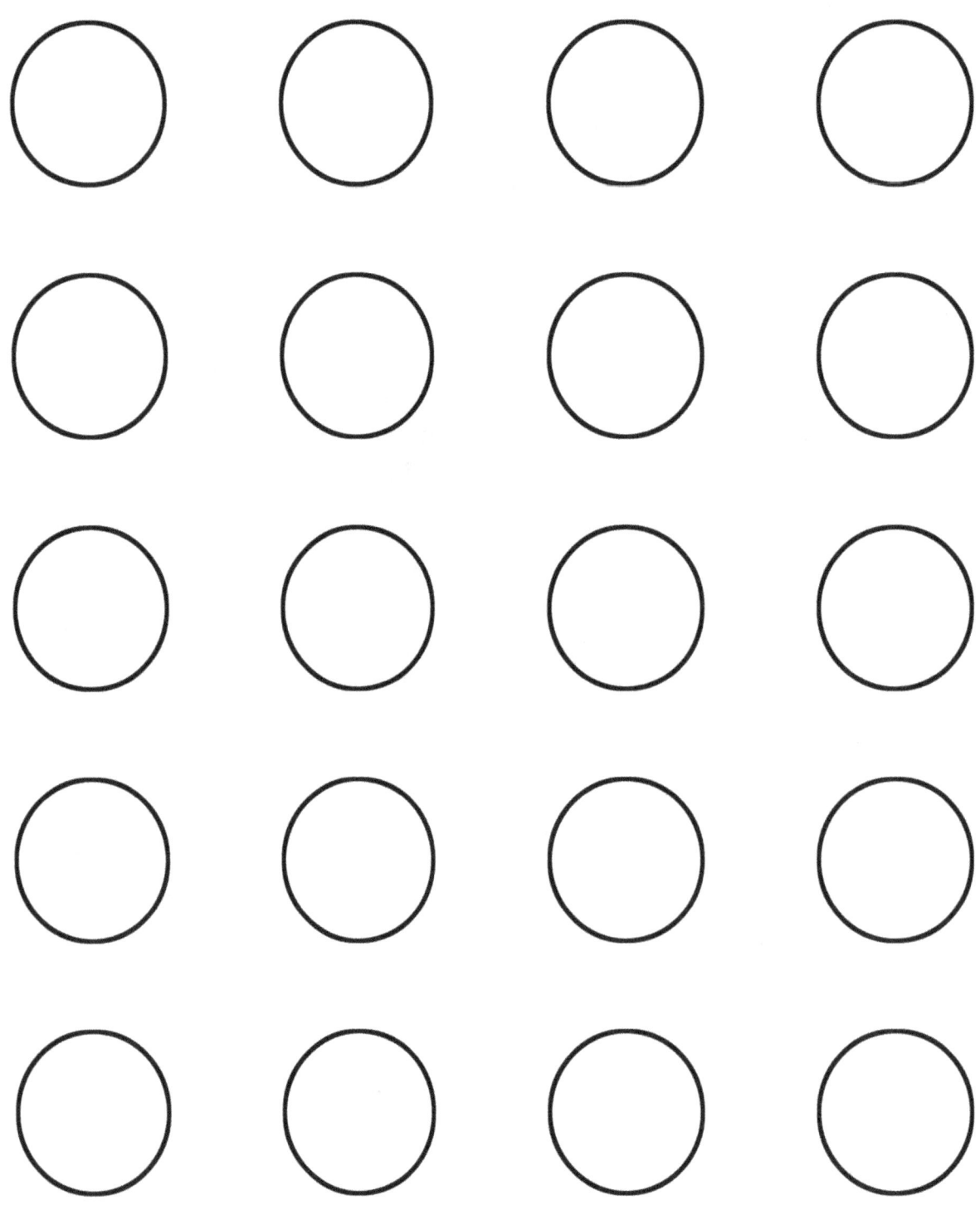

1 blue   3 dark green   5 pink
2 light green 4 orange   6 nude

Test Your Color

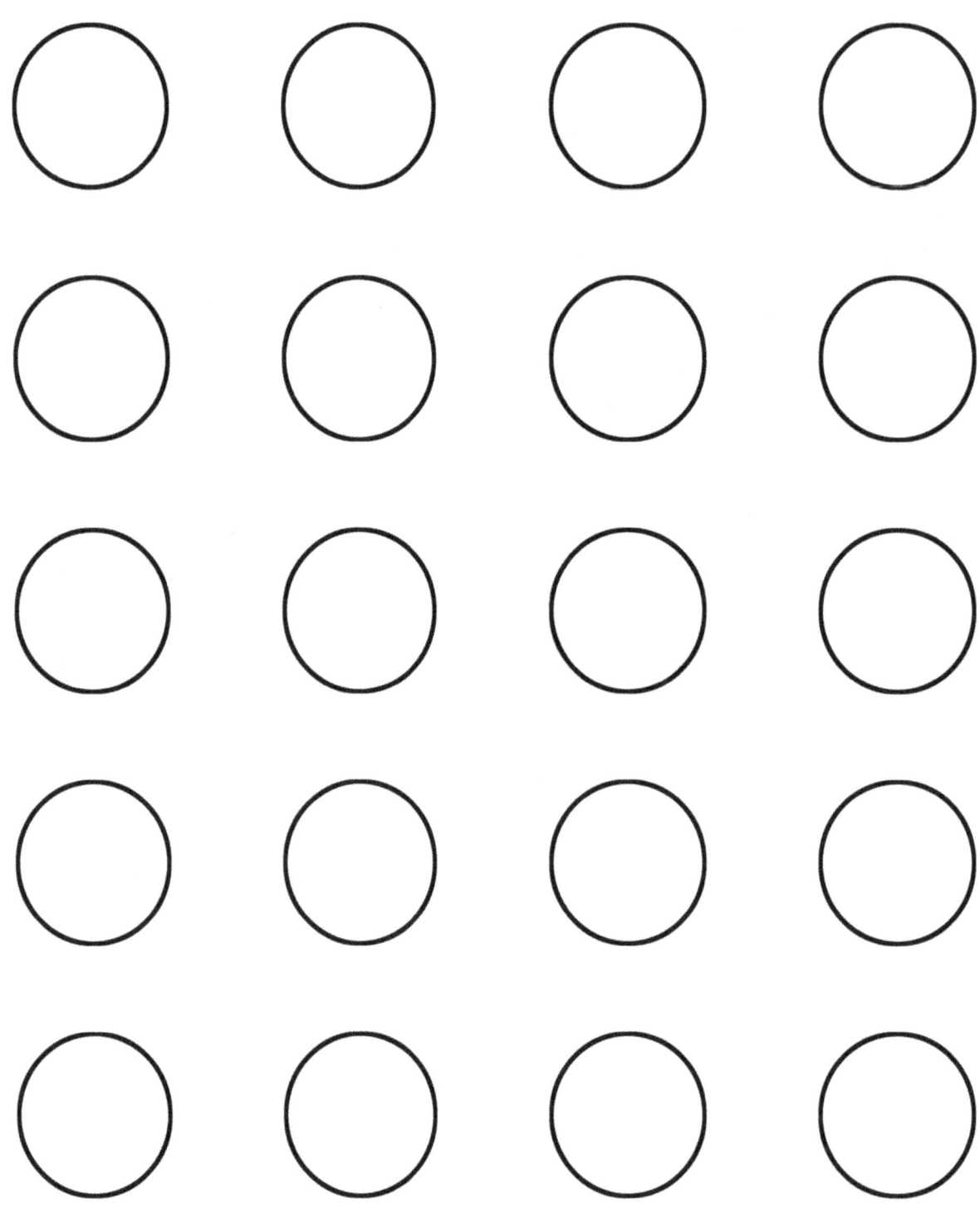

③ yellow ① blue
④ grey   ② pink

Test Your Color

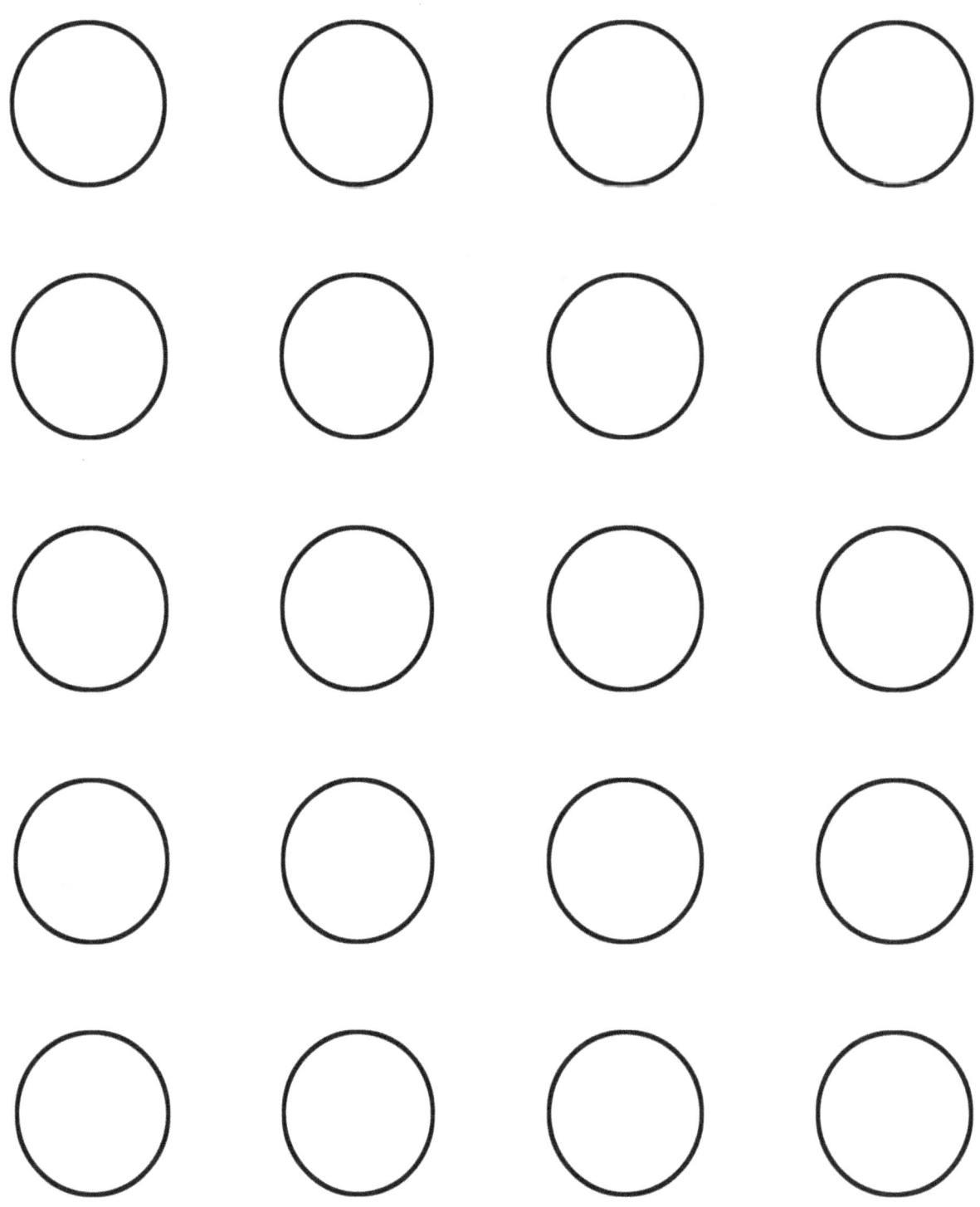

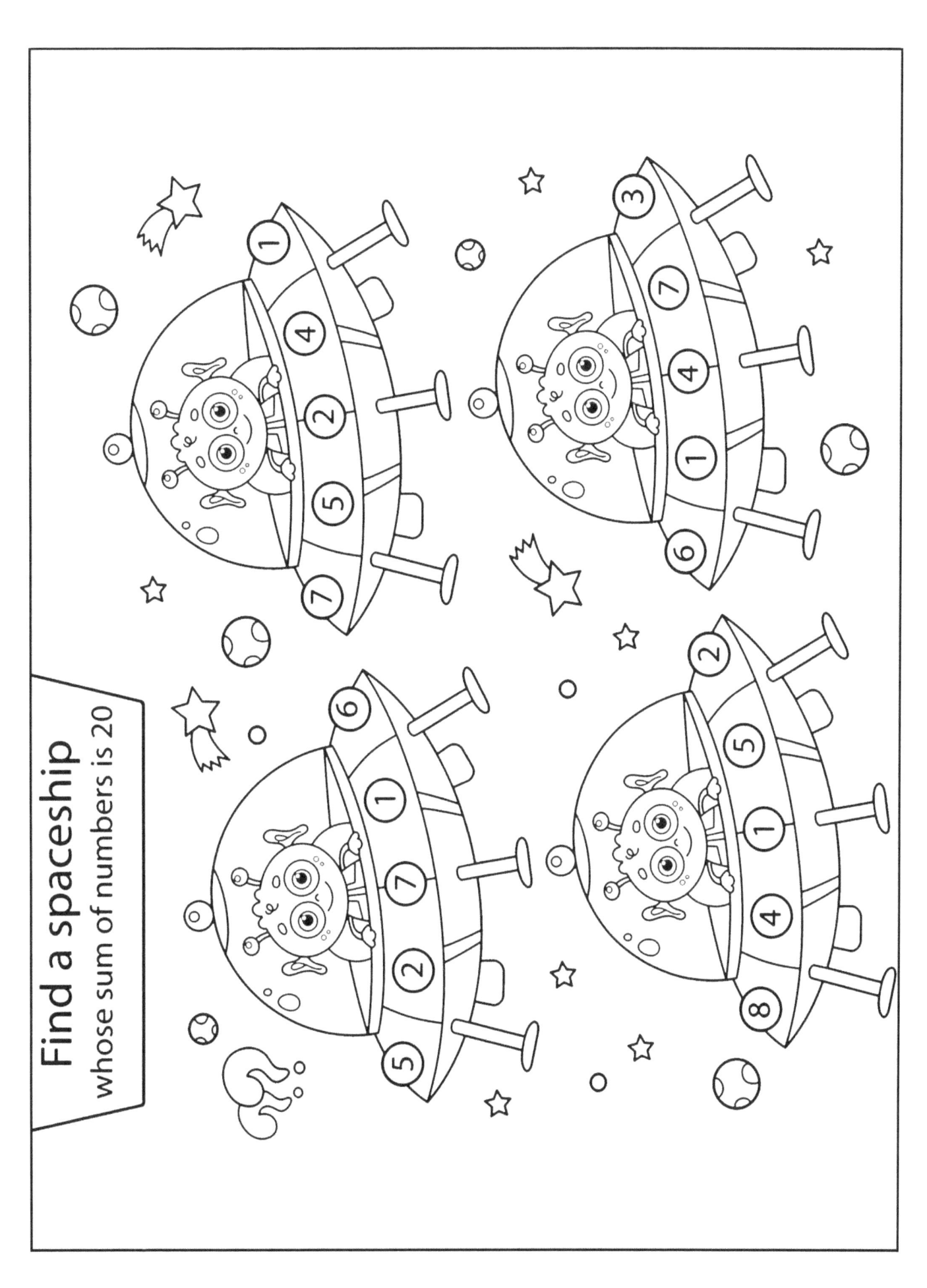

Find a spaceship
whose sum of numbers is 20

Test Your Color

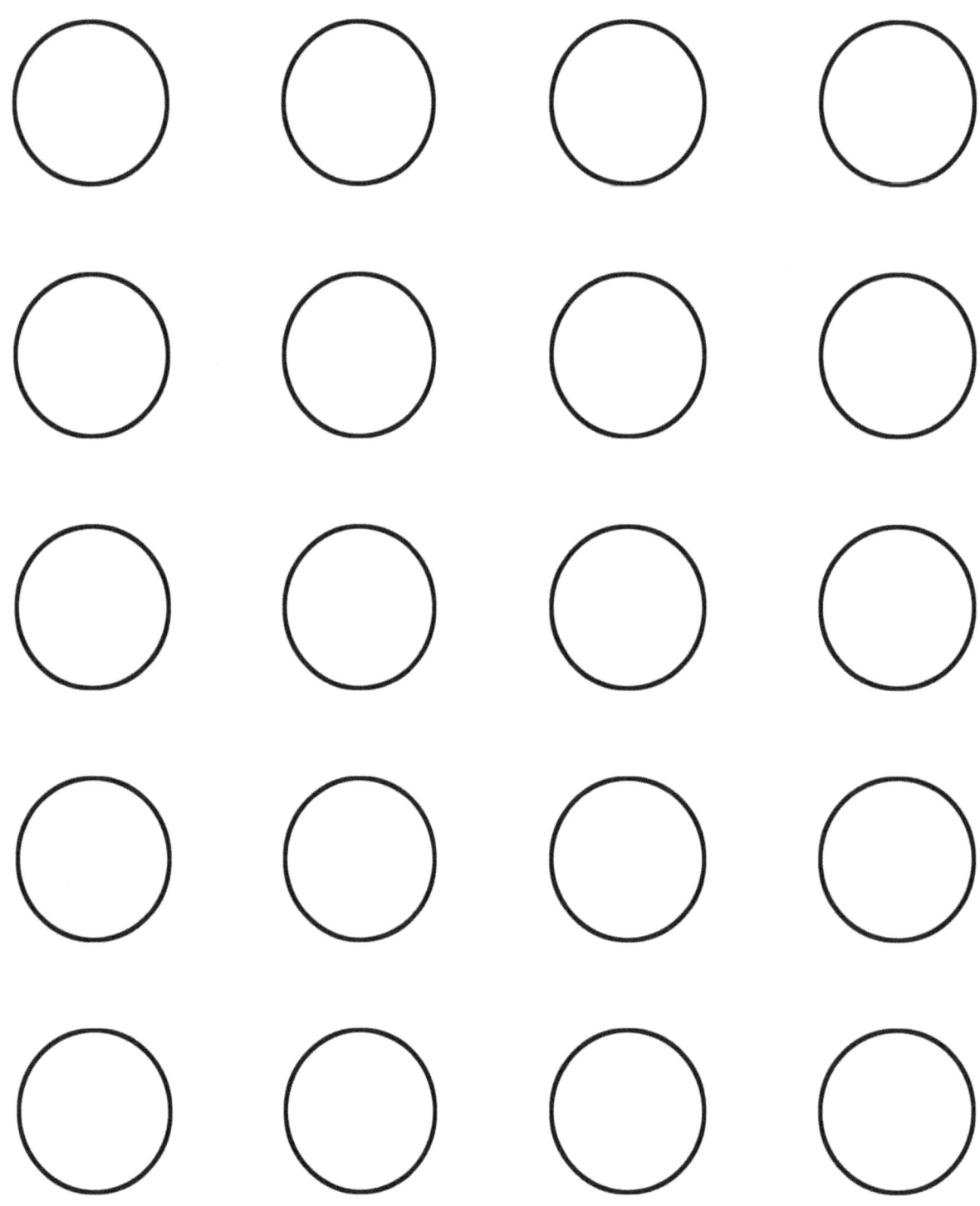

Which of the spaceship contains all these numbers?

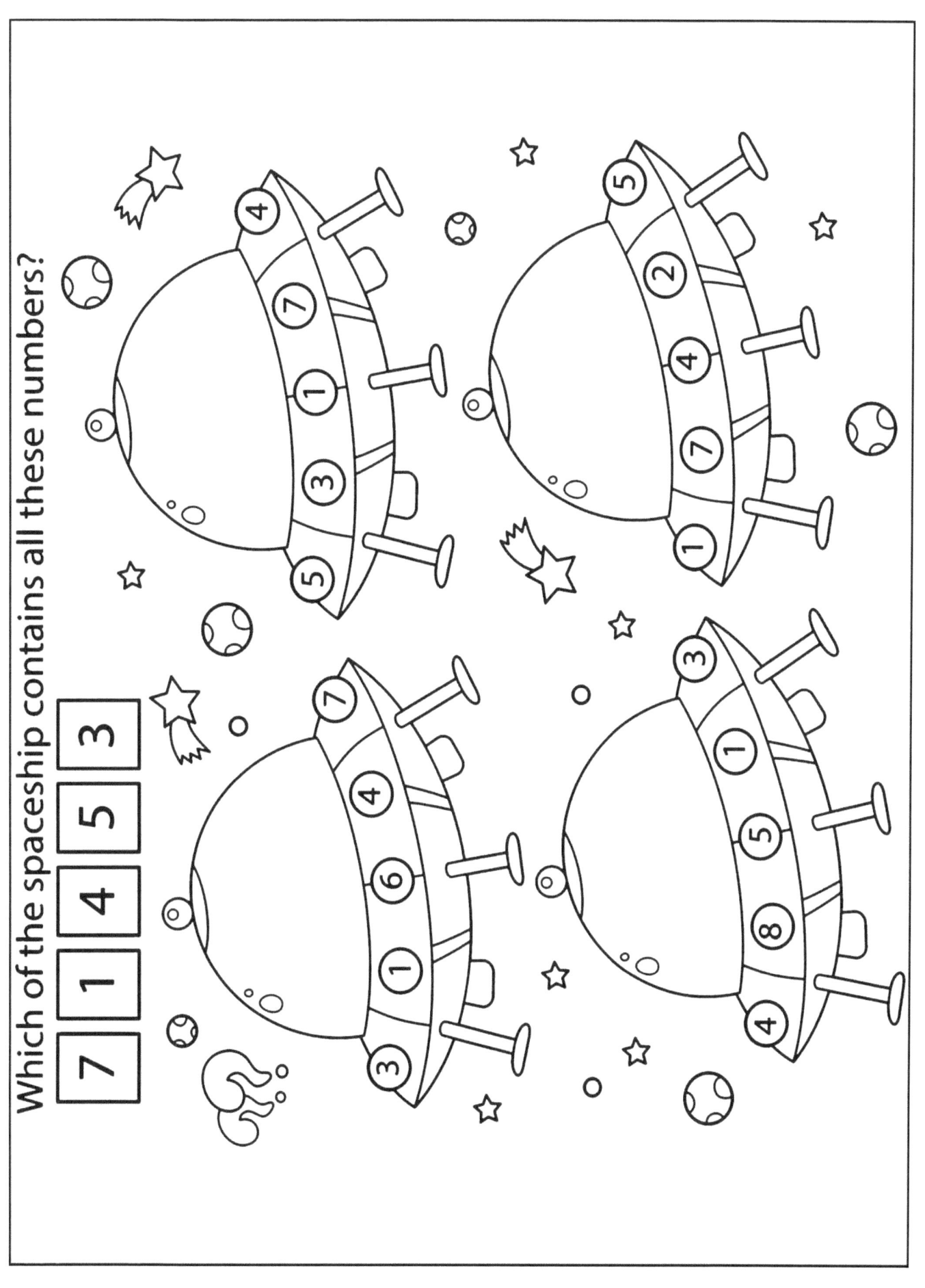

Test Your Color

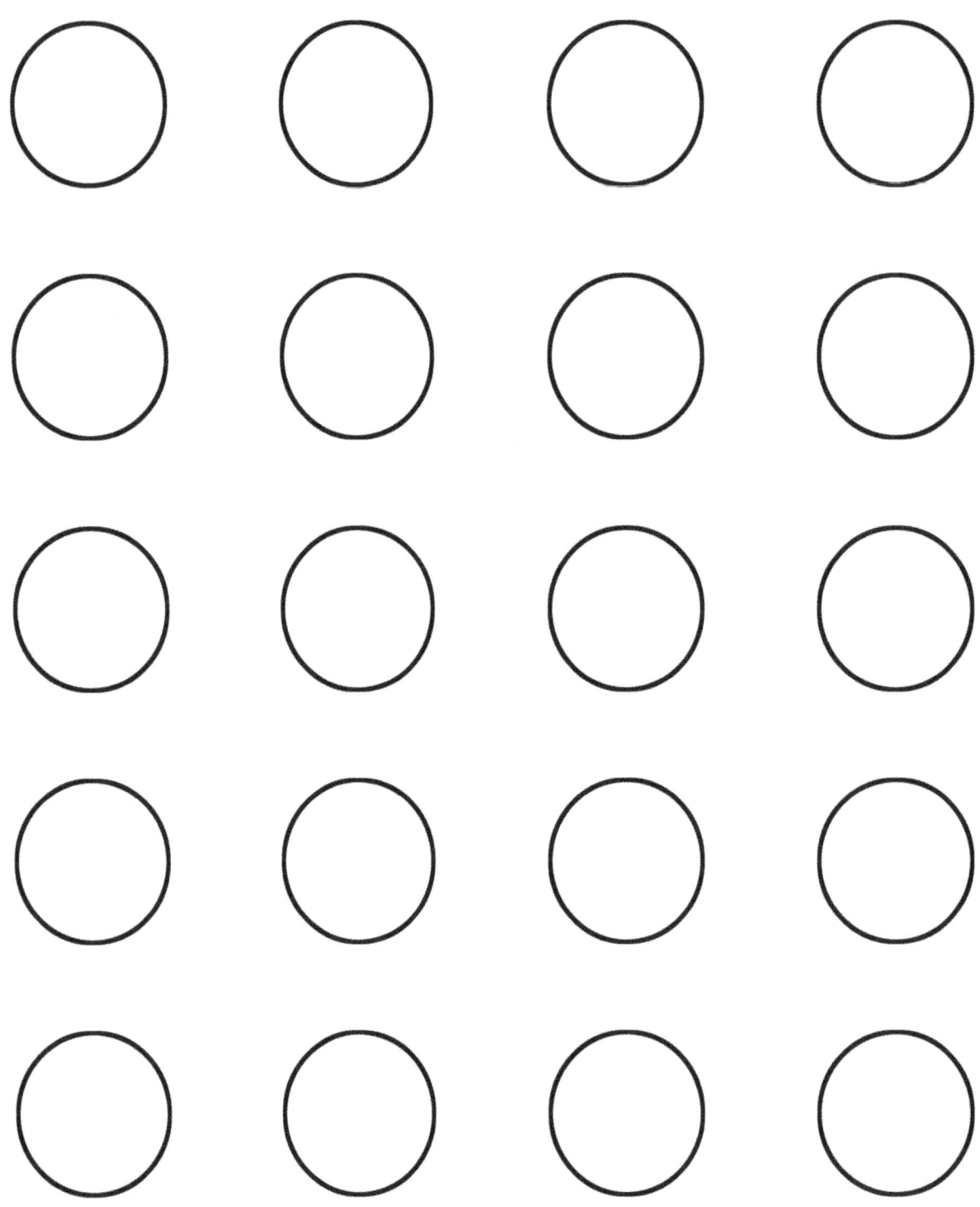

Connect the dots

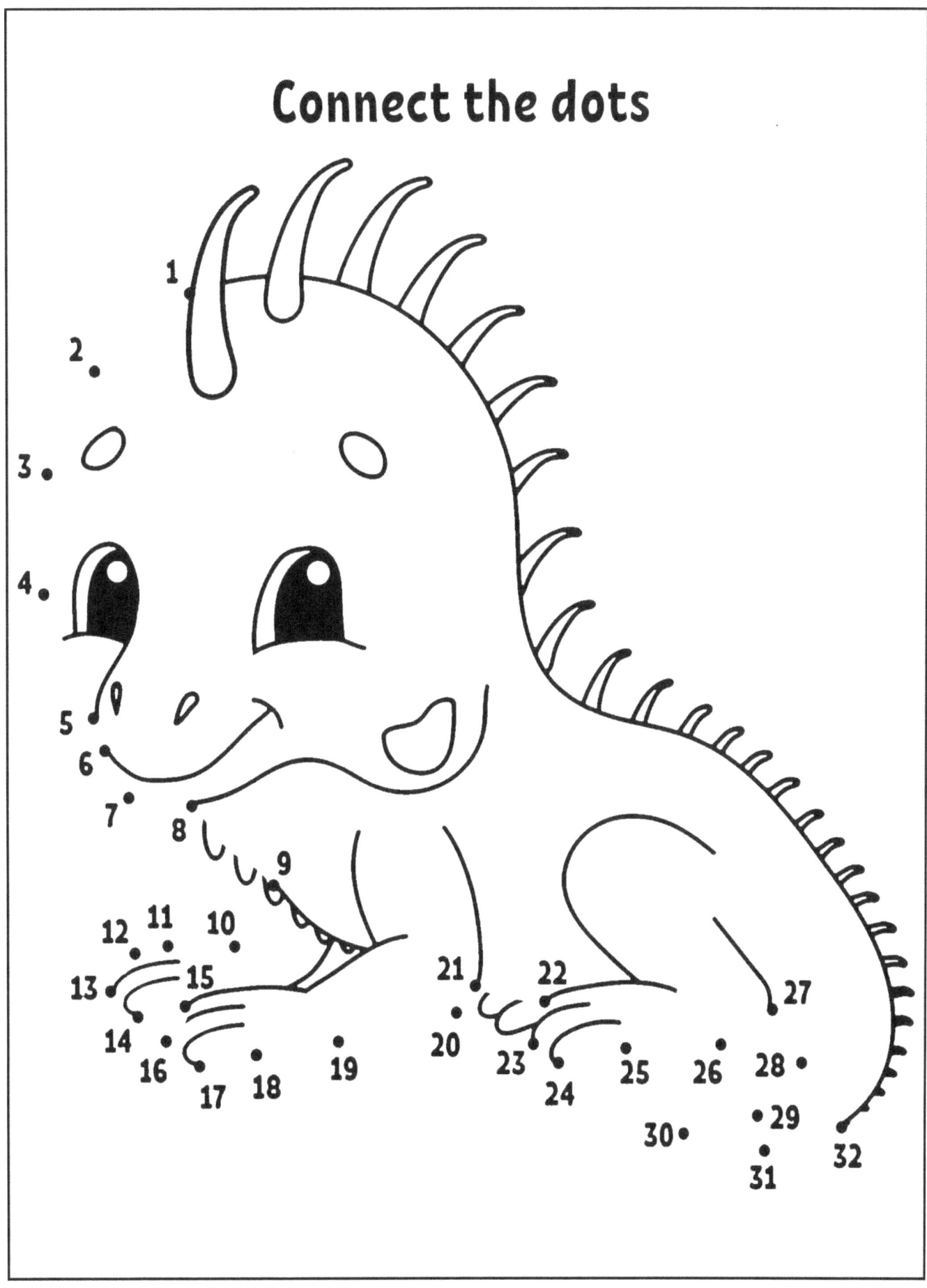

Test Your Color

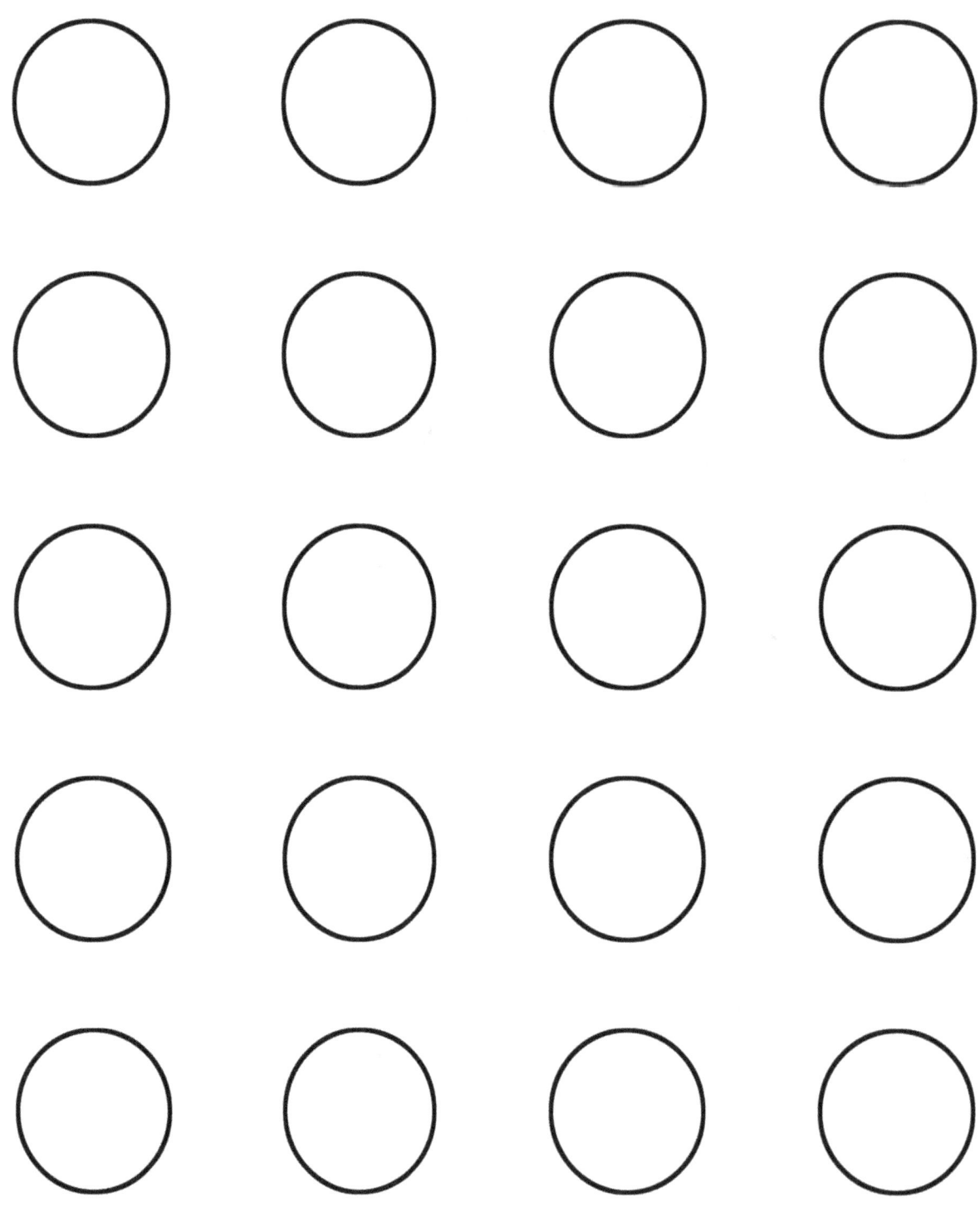

Connect the dots

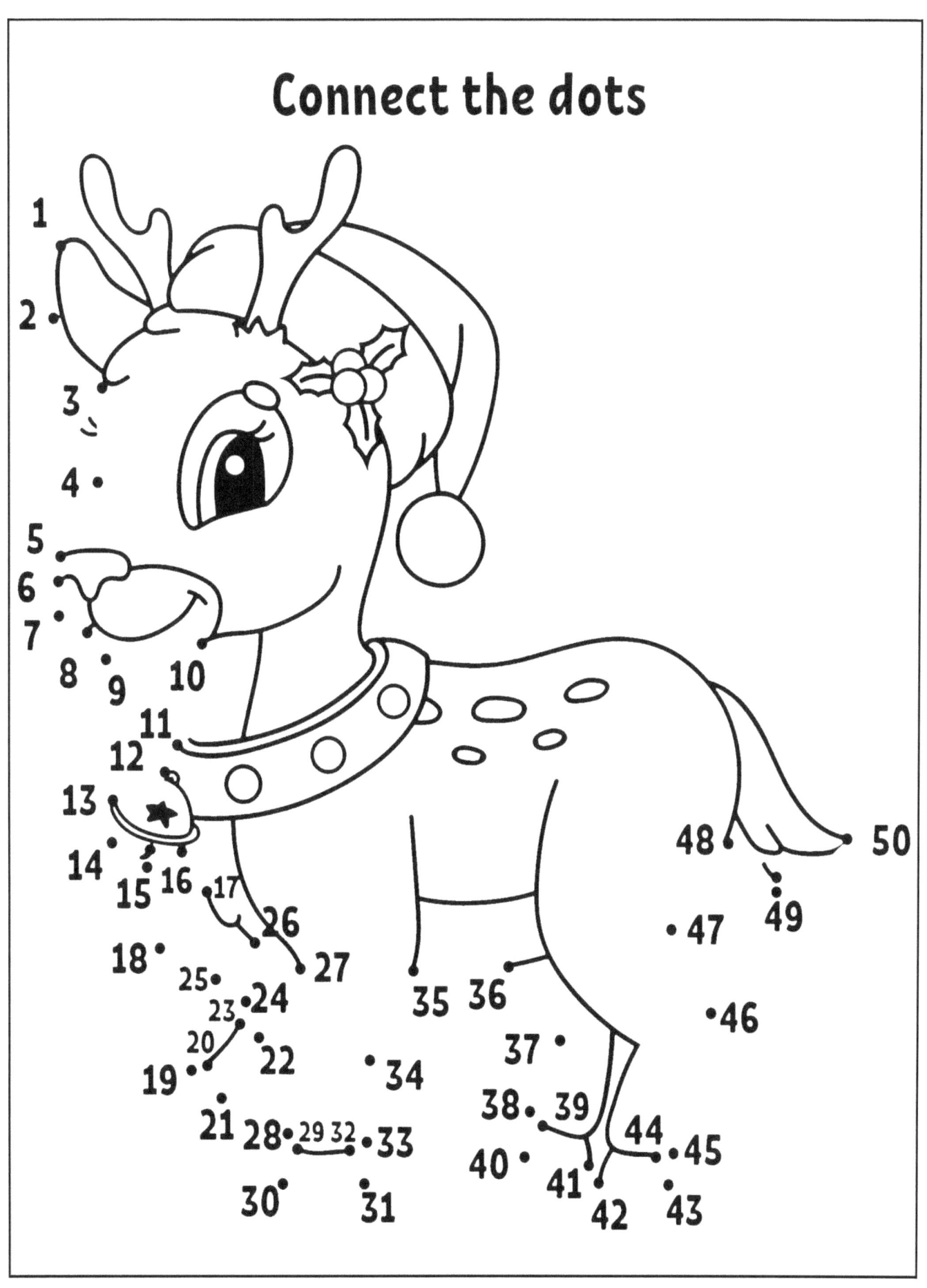

Test Your Color

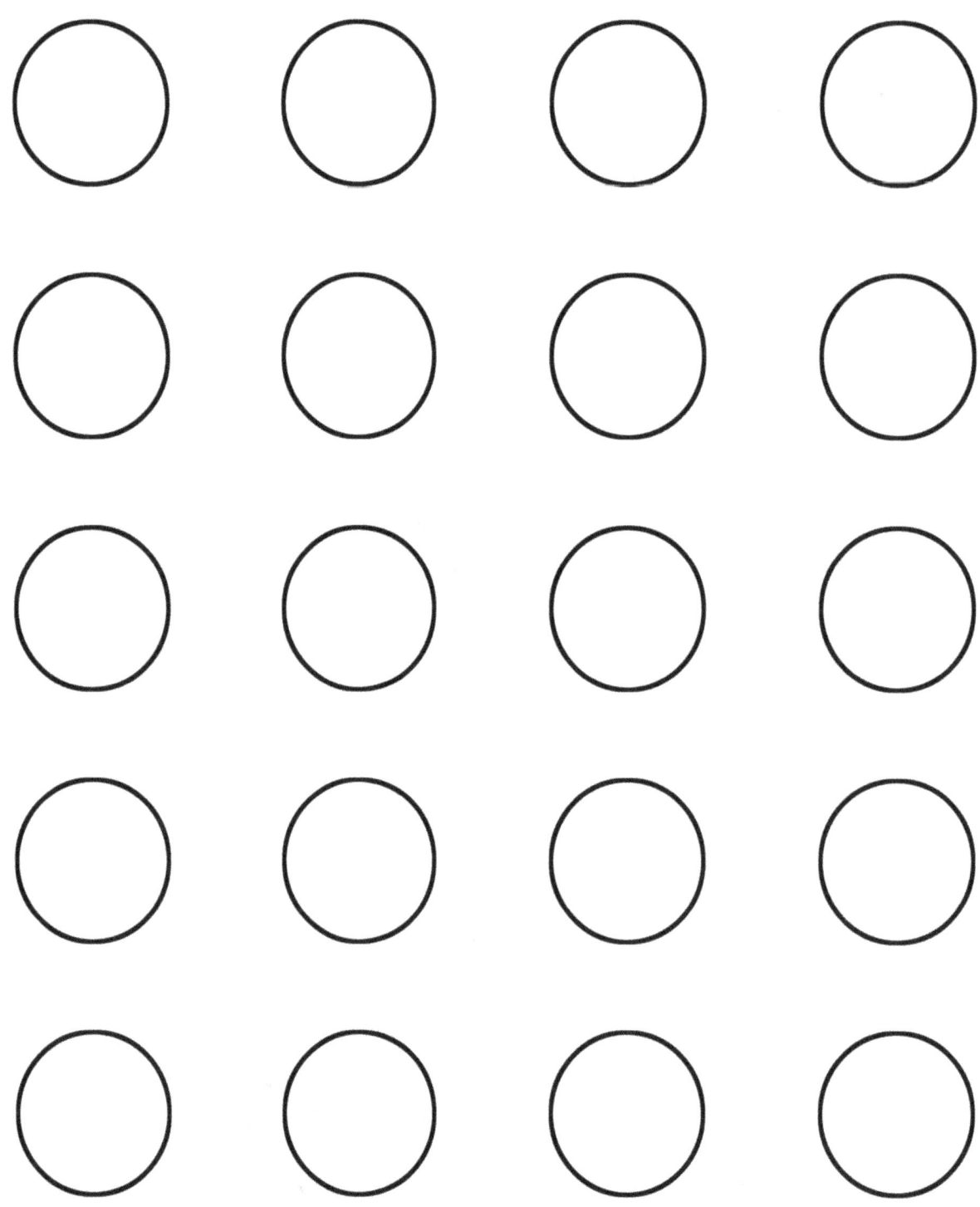

Connect the dots

Test Your Color

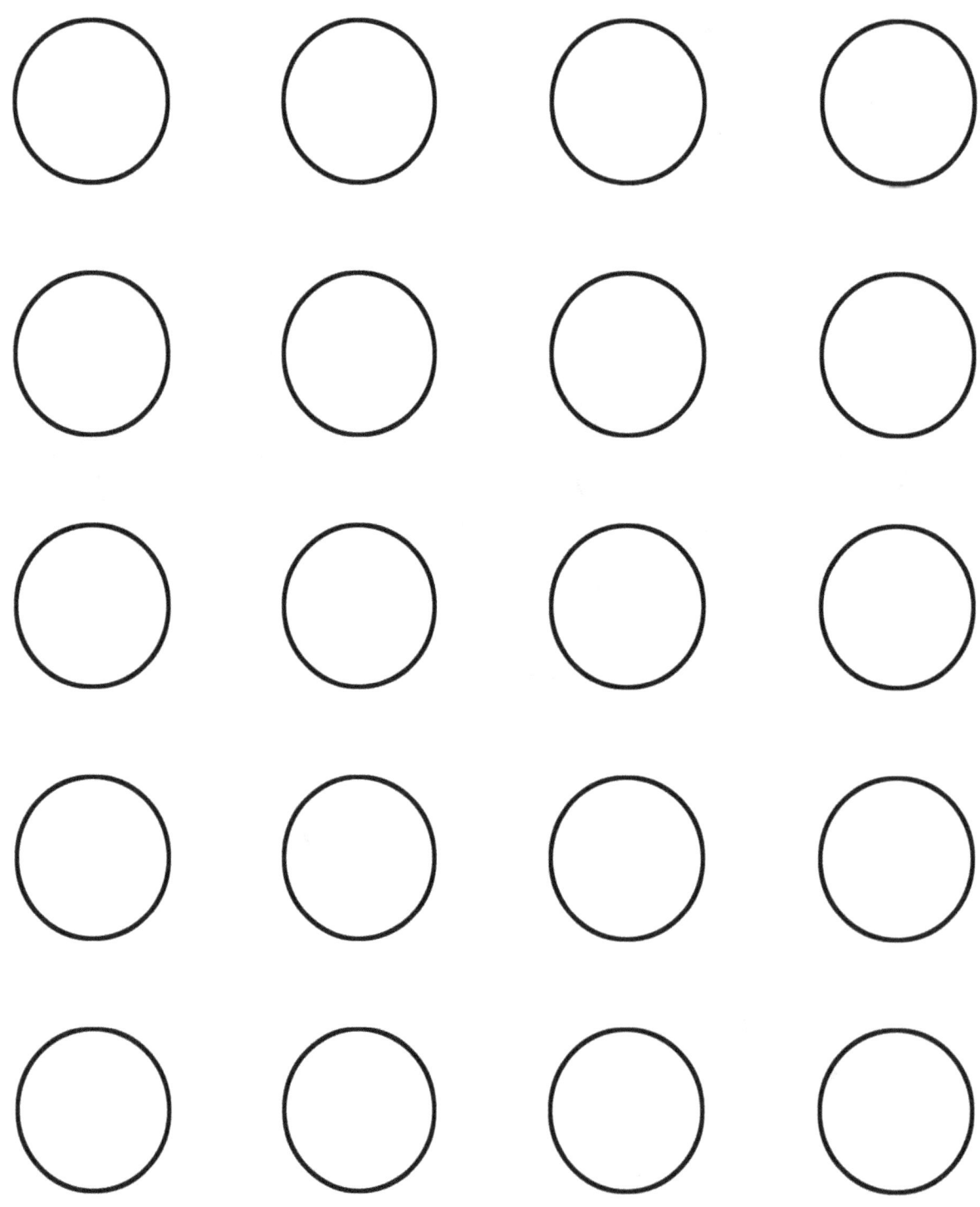

Draw a line from dot number 1 to dot number 2, then from dot number 2 to dot number 3, 3 to 4, and so on. Continue to join the dots until you have connected all the numbered dots. Then, color the picture!

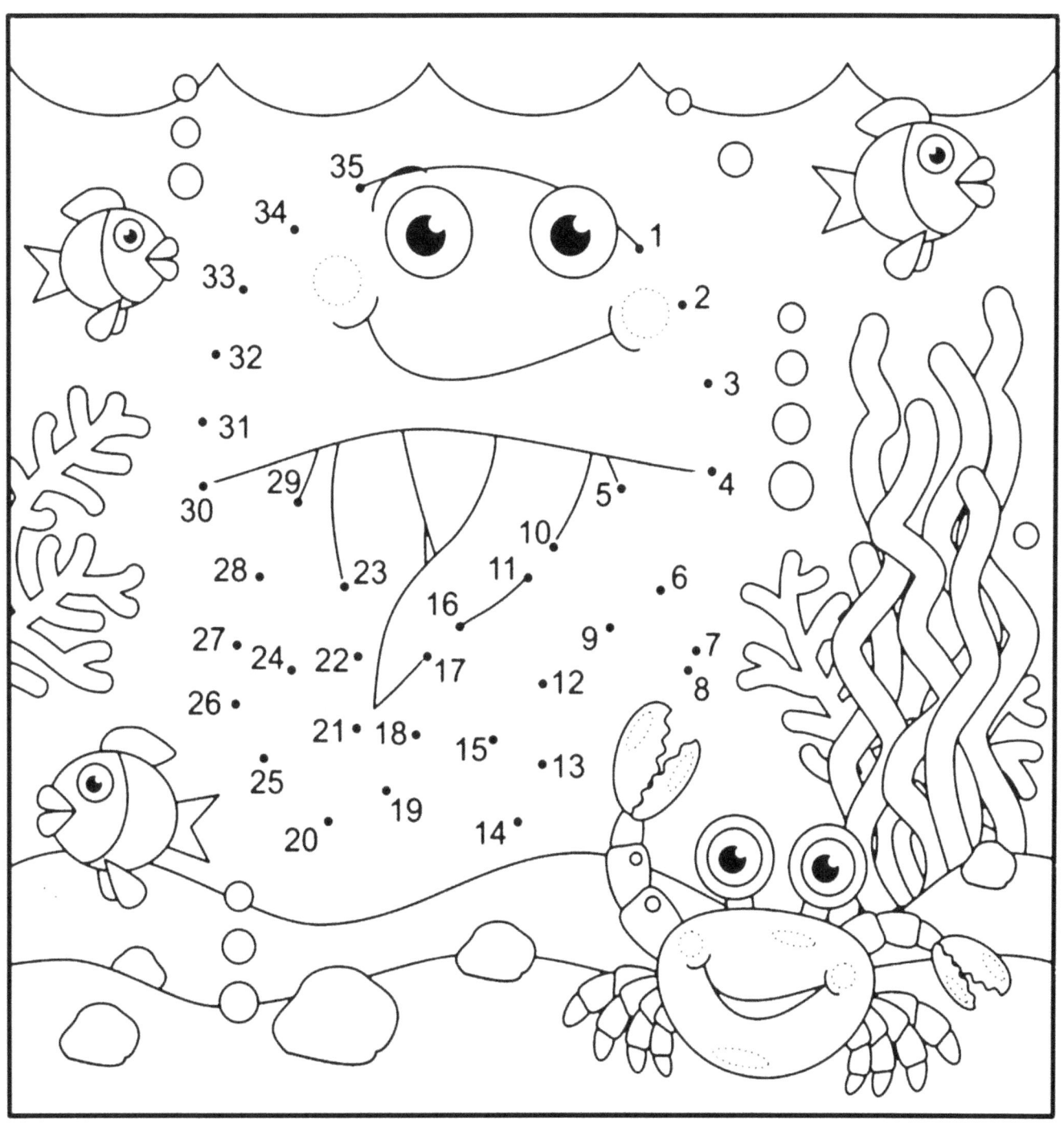

Test Your Color

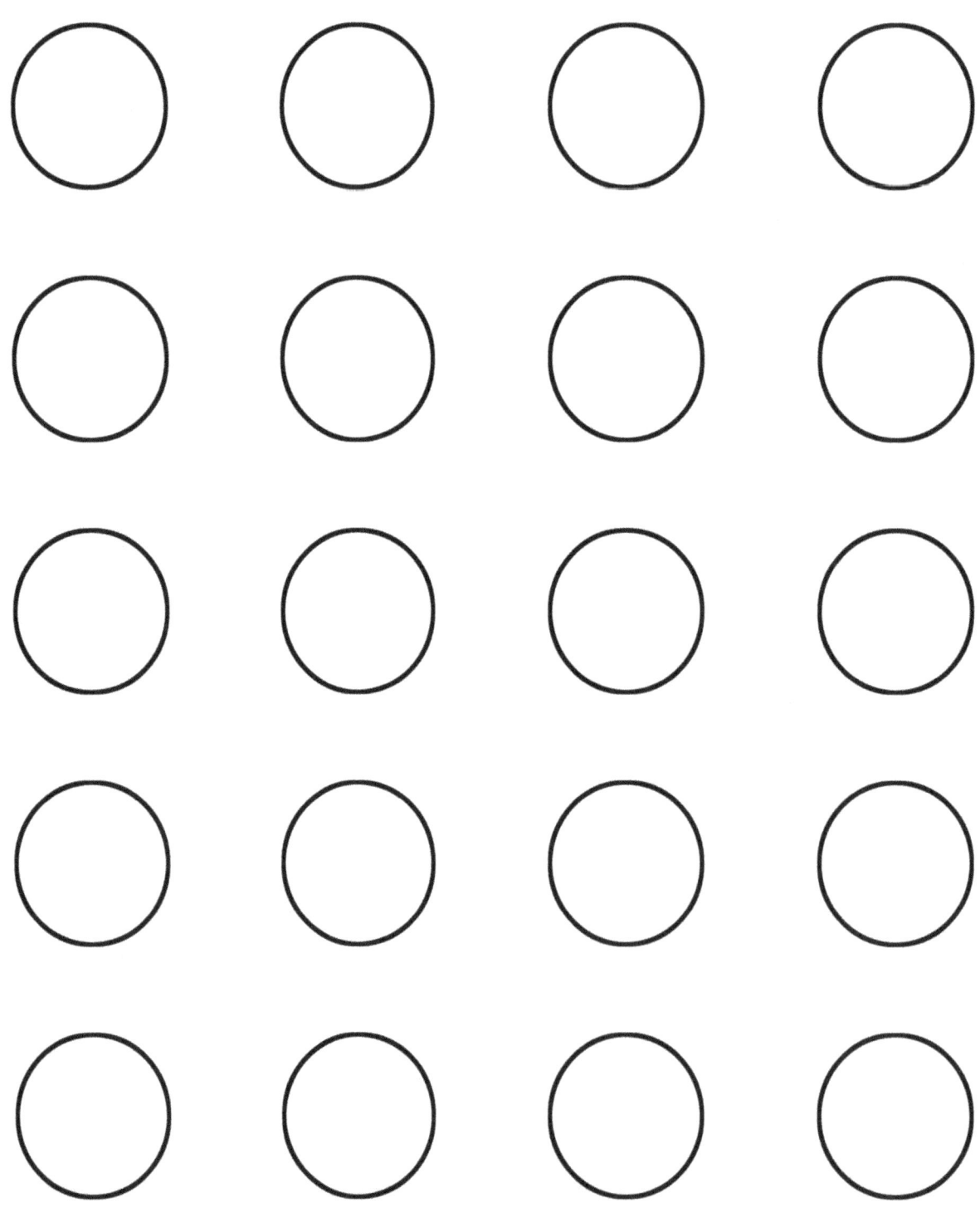

Draw a line from dot number 1 to dot number 2, then from dot number 2 to dot number 3, 3 to 4, and so on. Continue to join the dots until you have connected all the numbered dots. Then, color the picture!

Test Your Color

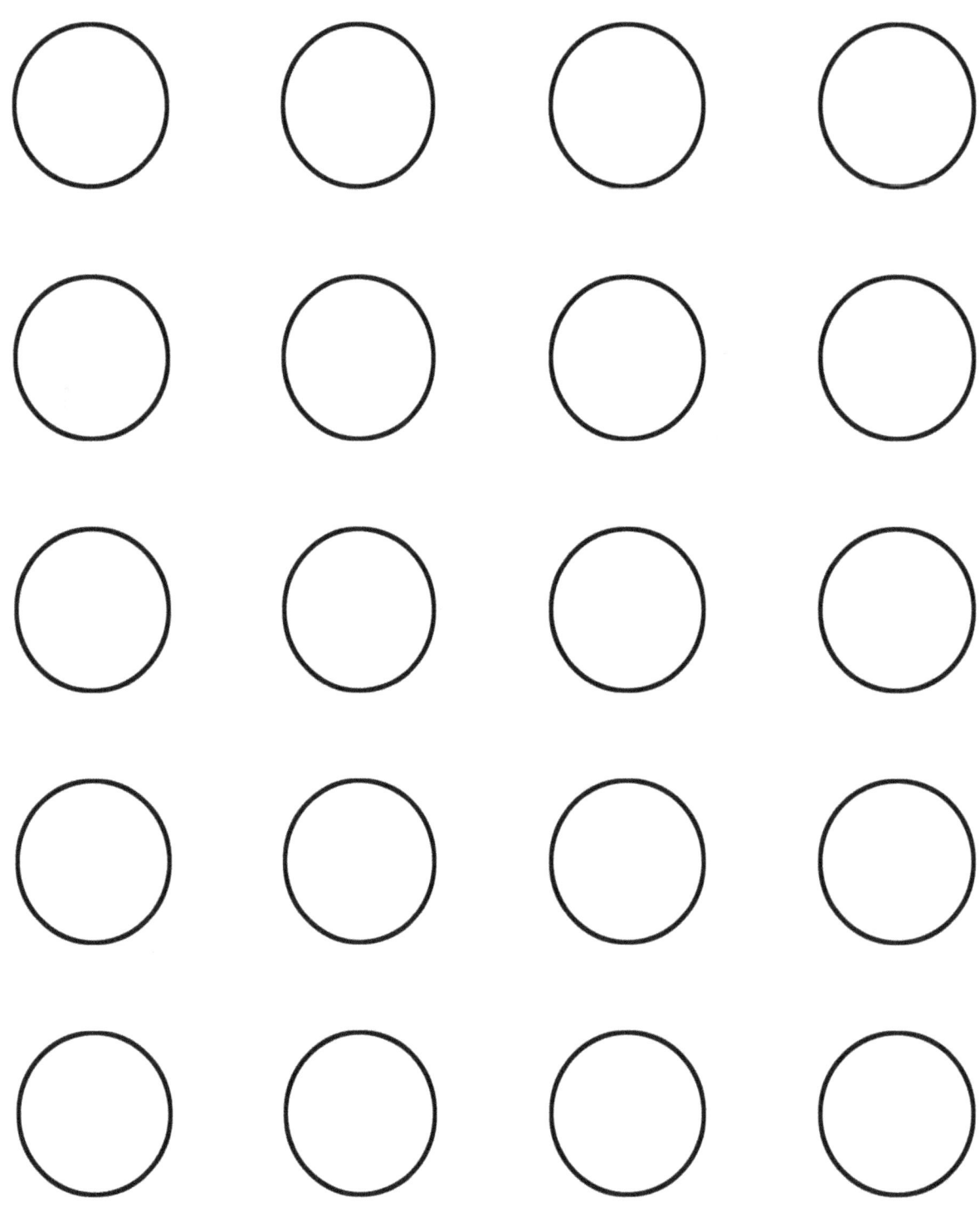

Draw a line from dot number 1 to dot number 2, then from dot number 2 to dot number 3, 3 to 4, and so on. Continue to join the dots until you have connected all the numbered dots. Then, color the picture!

Test Your Color

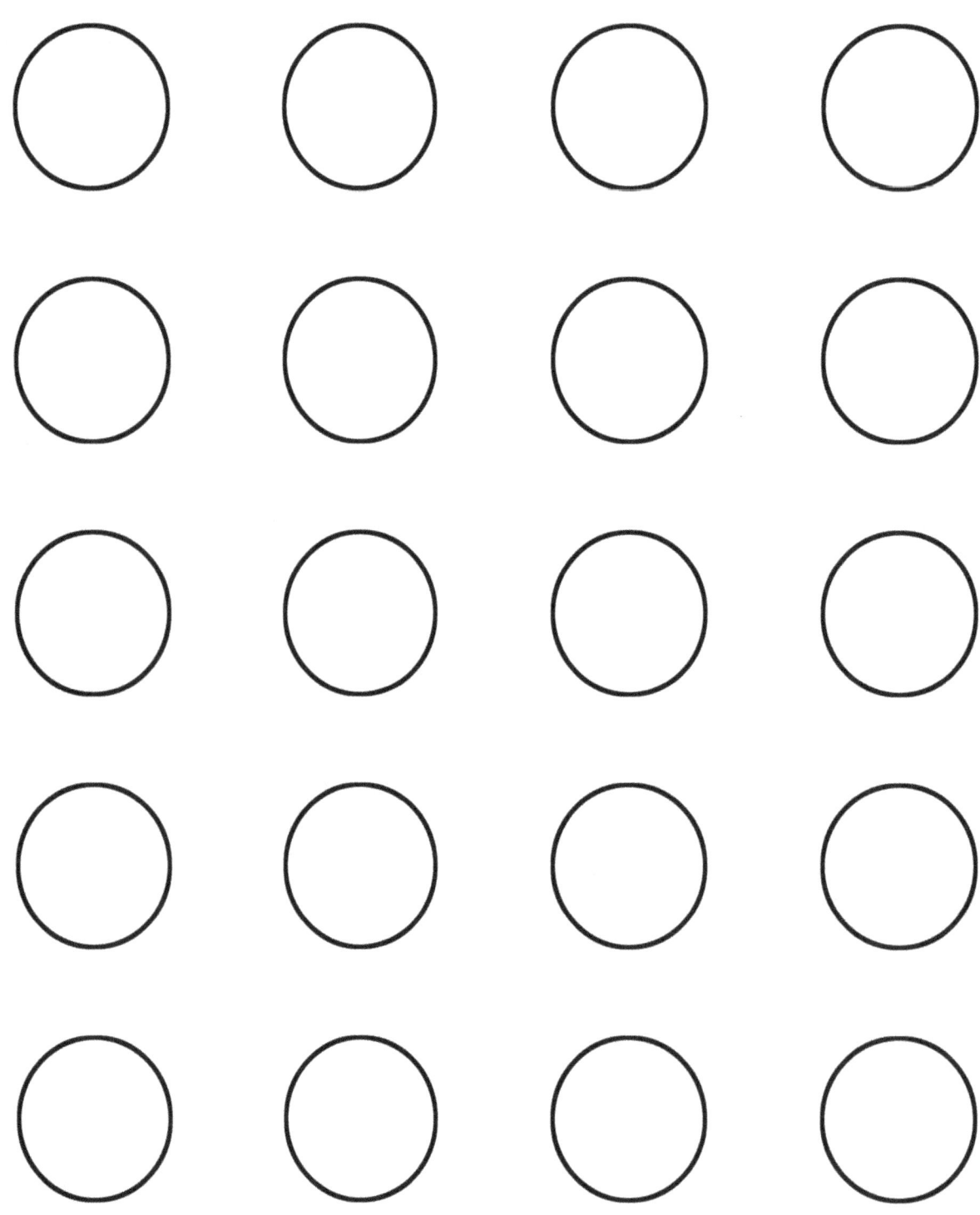

Draw a line from dot number 1 to dot number 2, then from dot number 2 to dot number 3, 3 to 4, and so on. Continue to join the dots until you have connected all the numbered dots. Then, color the picture!

Test Your Color

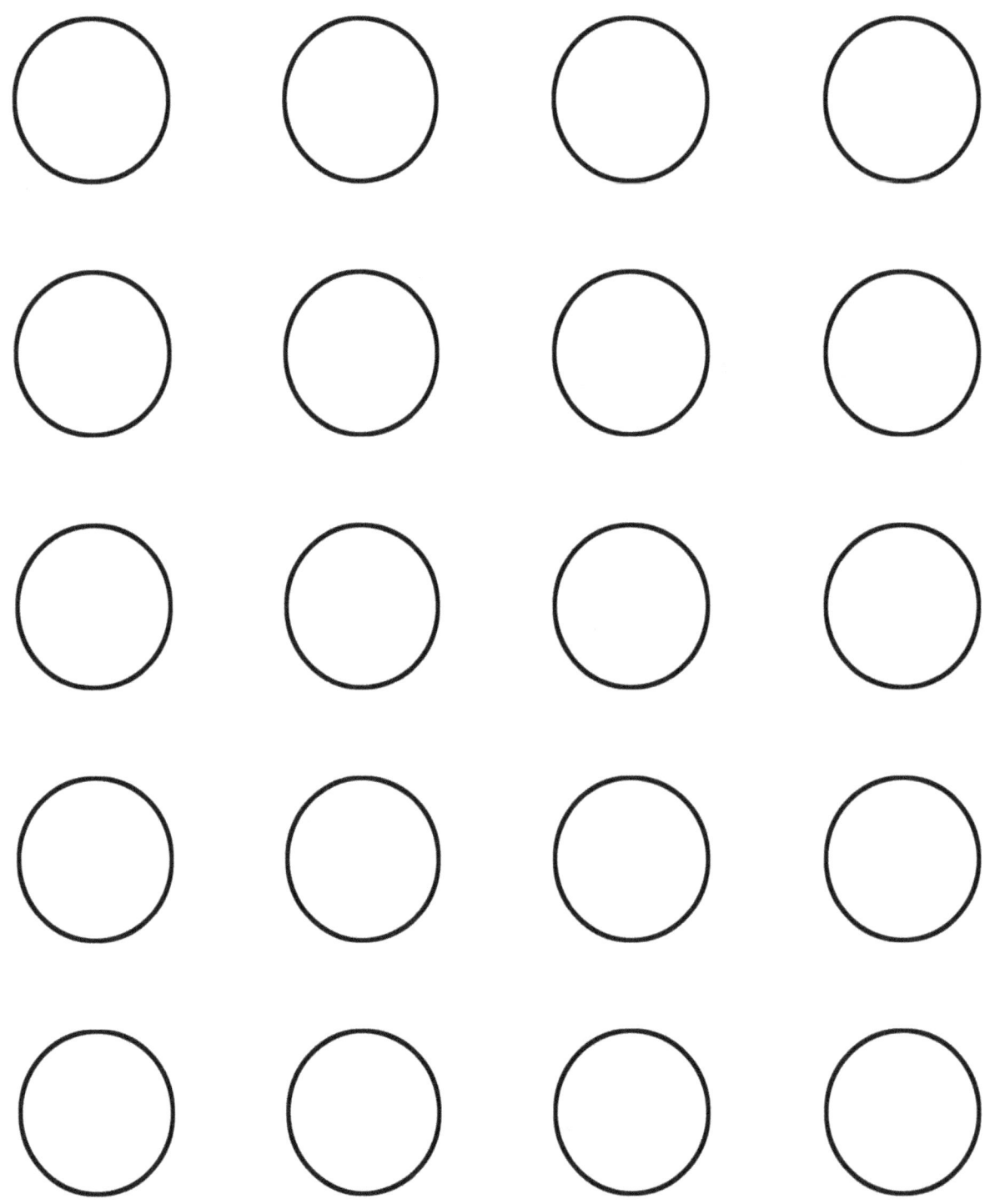

Draw a line from dot number 1 to dot number 2, then from dot number 2 to dot number 3, 3 to 4, and so on. Continue to join the dots until you have connected all the numbered dots. Then, color the picture!

Test Your Color

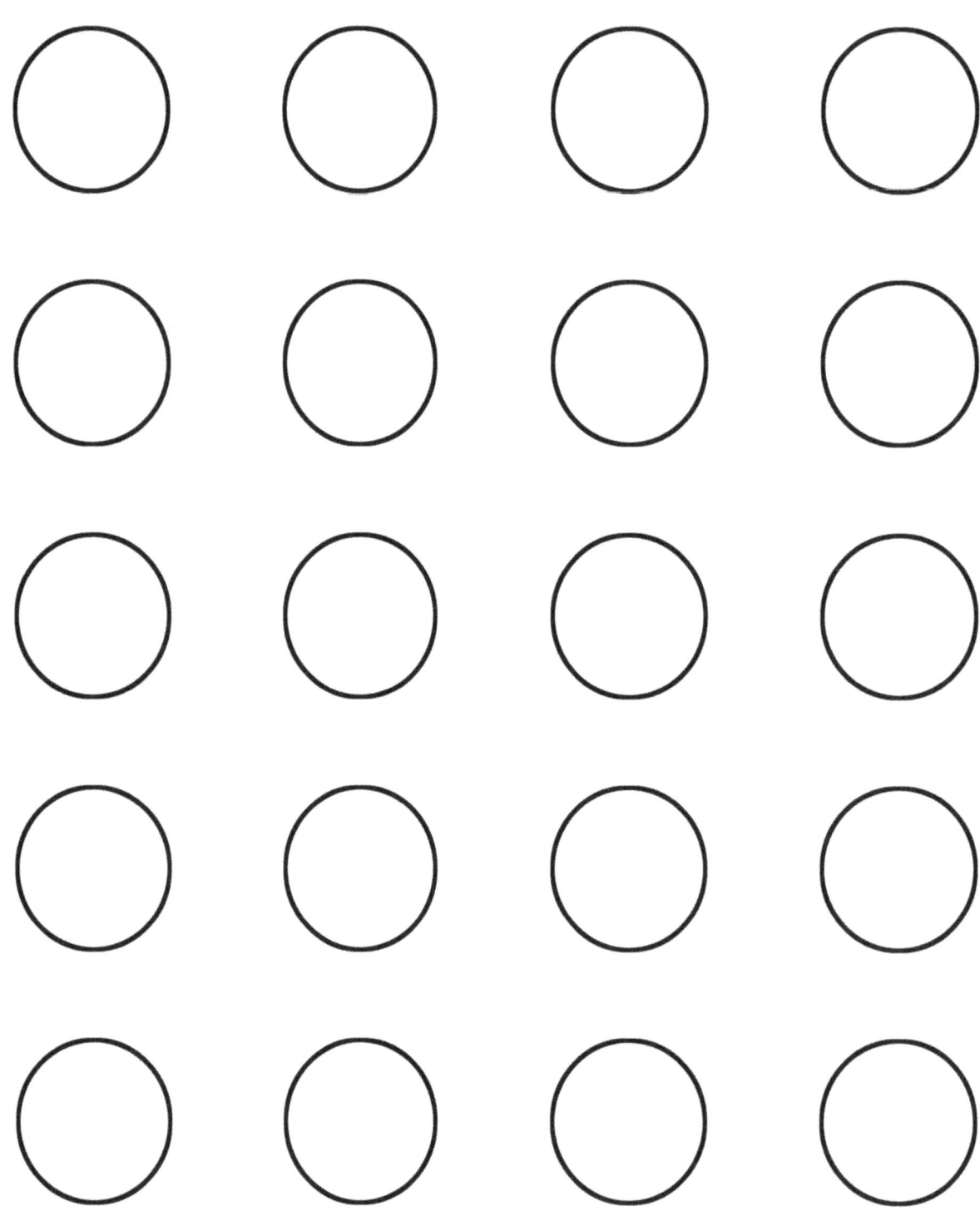

Draw a line from dot number 1 to dot number 2, then from dot number 2 to dot number 3, 3 to 4, and so on. Continue to join the dots until you have connected all the numbered dots. Then, color the picture!

Test Your Color

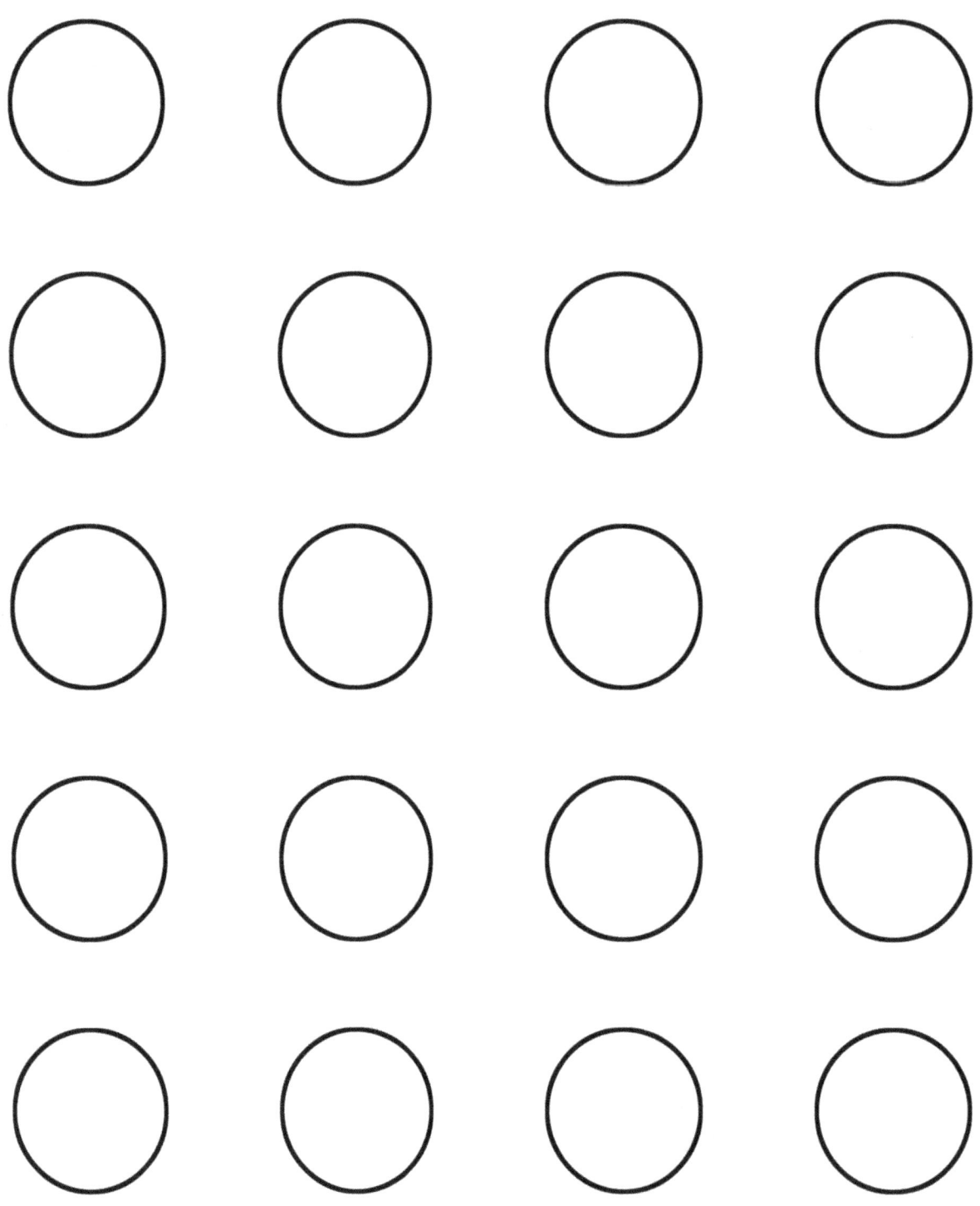

Draw a line from dot number 1 to dot number 2, then from dot number 2 to dot number 3, 3 to 4, and so on. Continue to join the dots until you have connected all the numbered dots. Then, color the picture!

Test Your Color

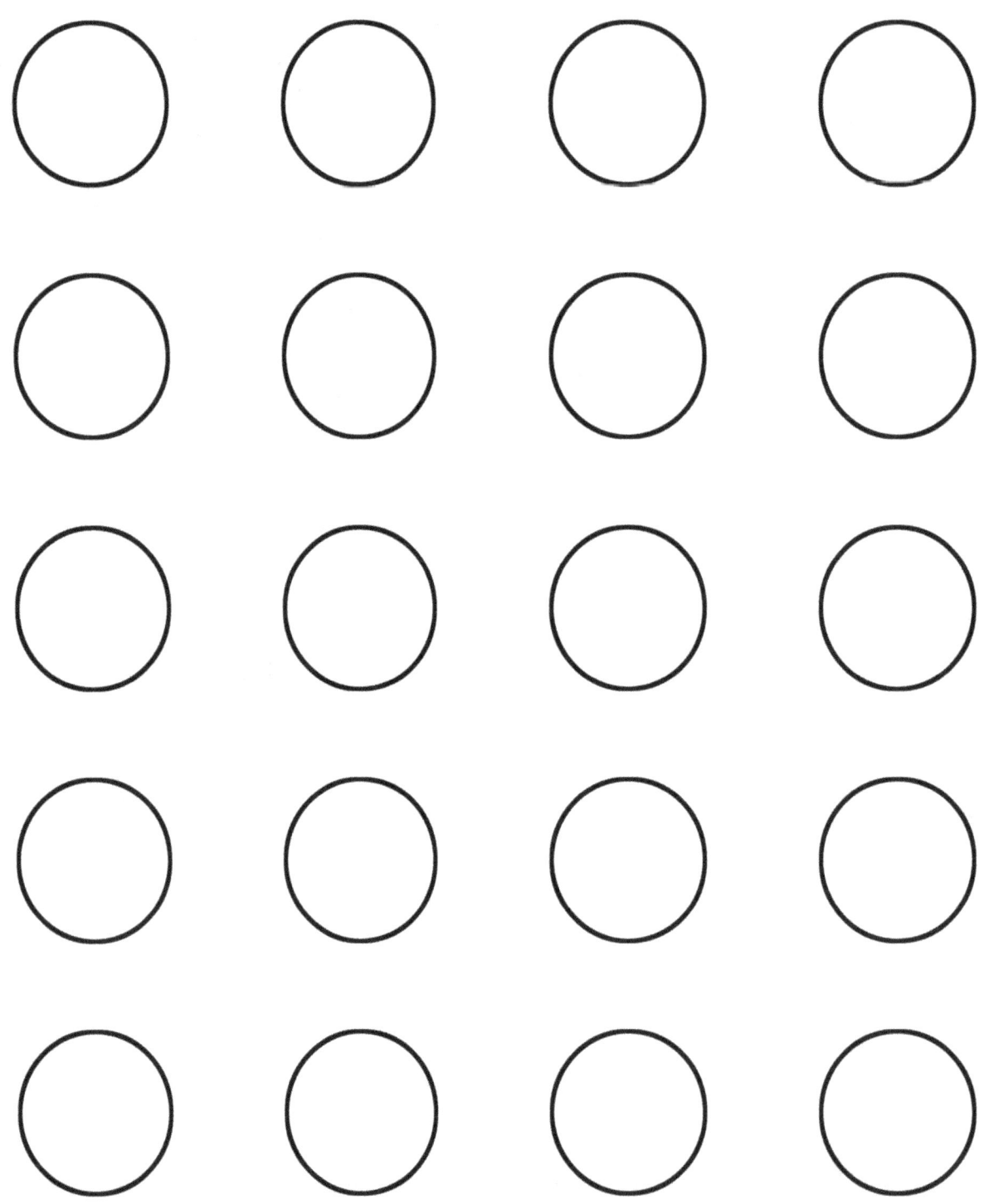

Draw a line from dot number 1 to dot number 2, then from dot number 2 to dot number 3, 3 to 4, and so on. Continue to join the dots until you have connected all the numbered dots. Then, color the picture!

Test Your Color

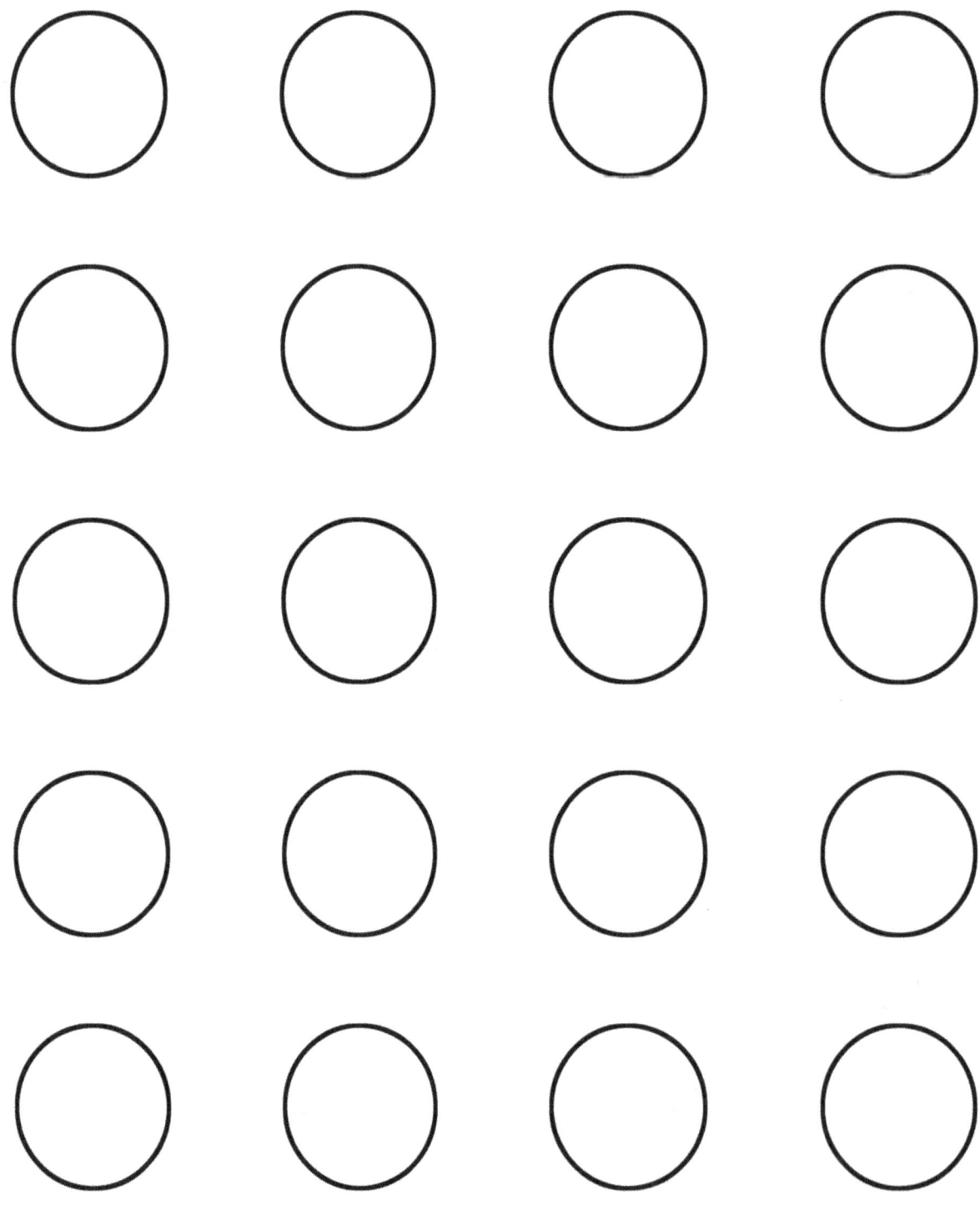

Draw a line from dot number 1 to dot number 2, then from dot number 2 to dot number 3, 3 to 4, and so on. Continue to join the dots until you have connected all the numbered dots. Then, color the picture!

Test Your Color

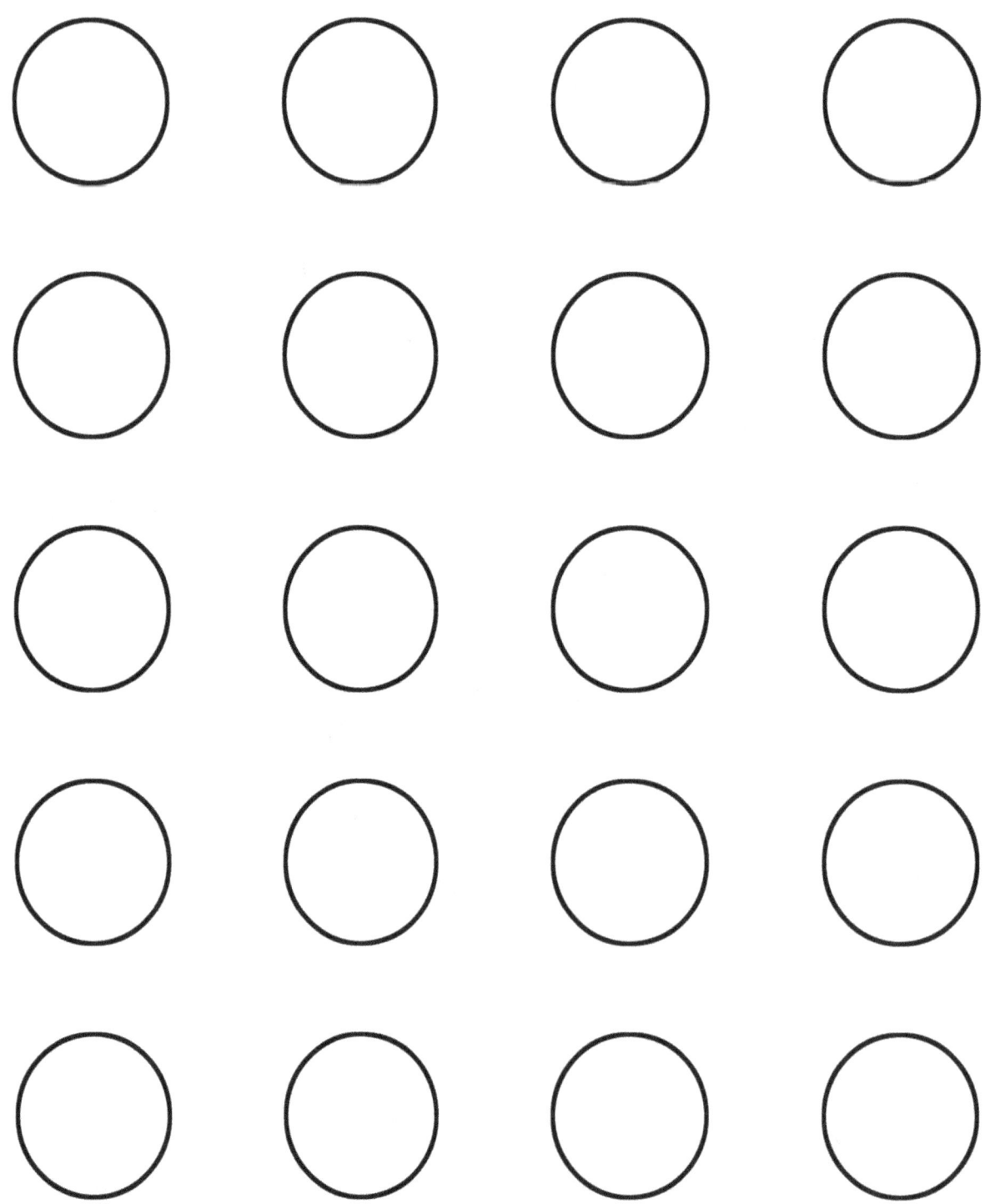

Draw a line from dot number 1 to dot number 2, then from dot number 2 to dot number 3, 3 to 4, and so on. Continue to join the dots until you have connected all the numbered dots. Then, color the picture!

Test Your Color

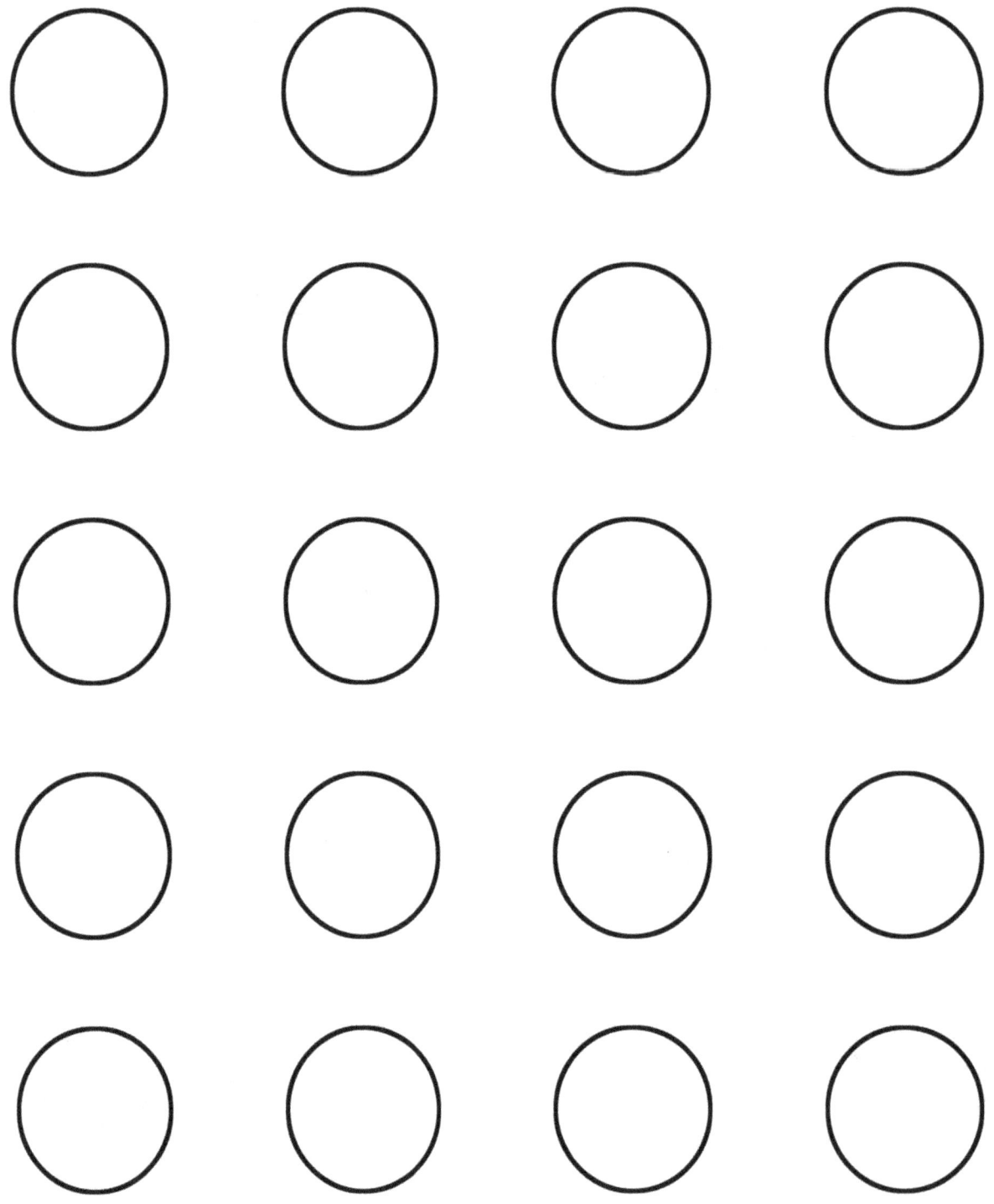

Connect the dots

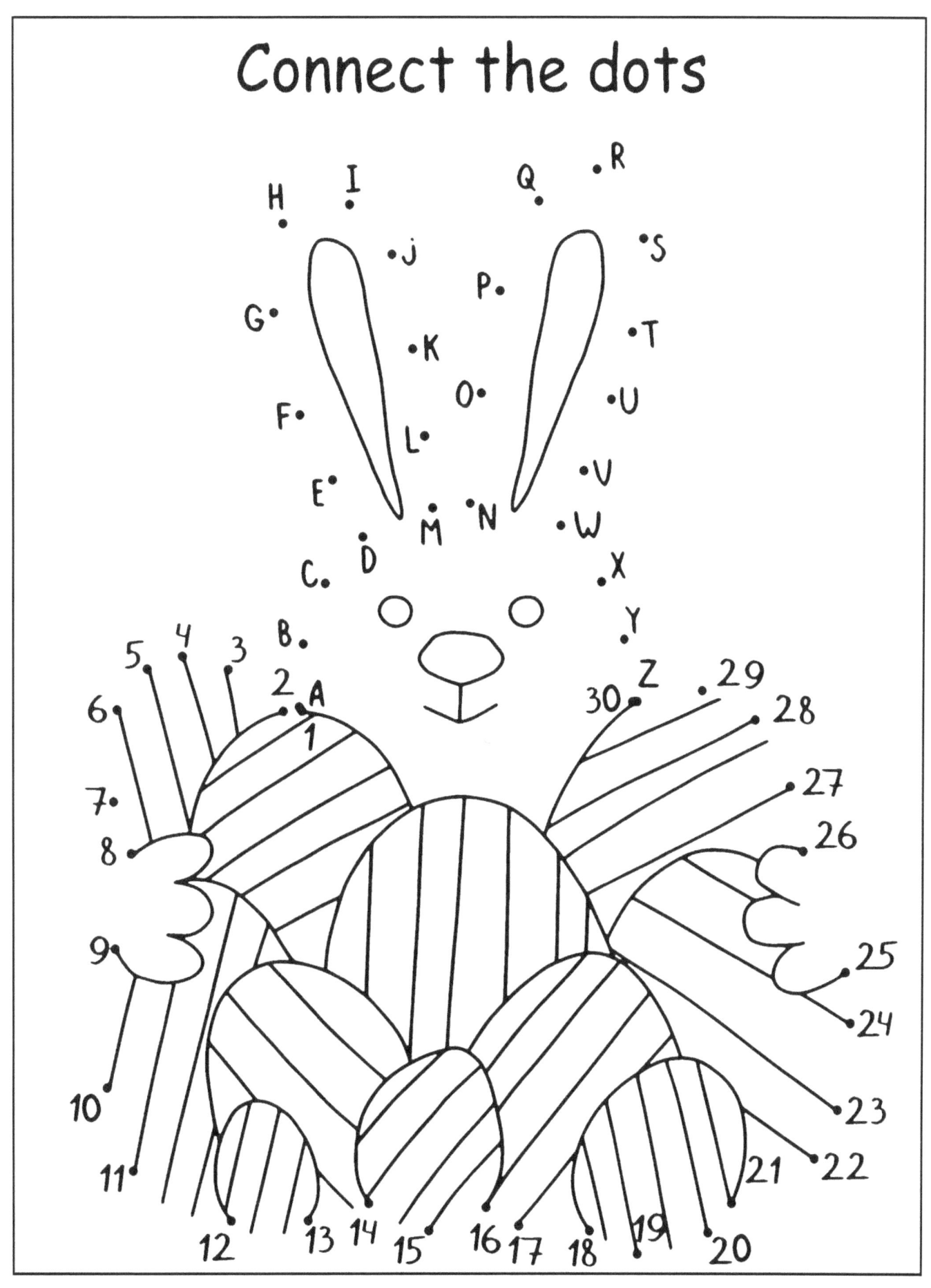

Test Your Color

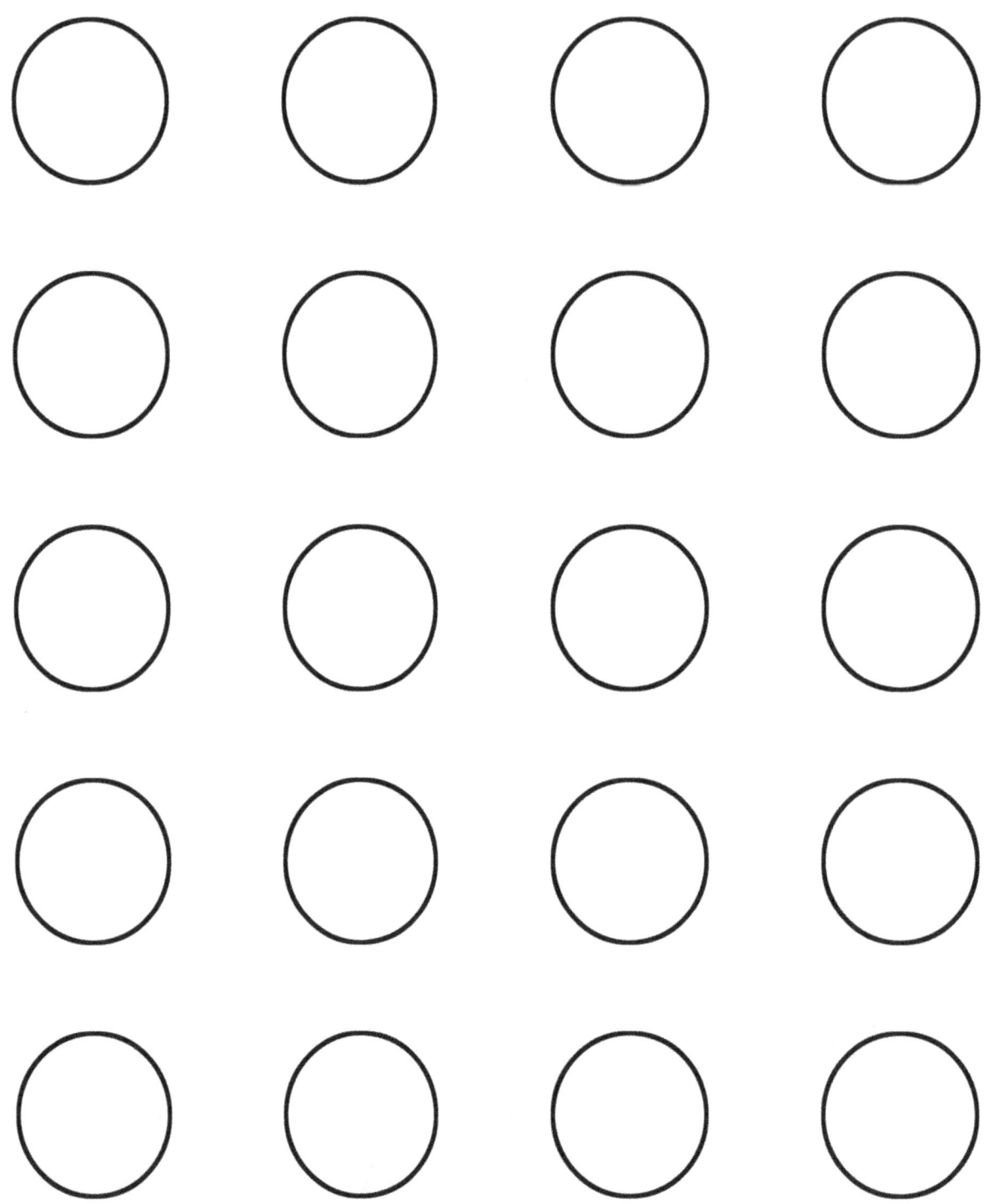

Connect the dots

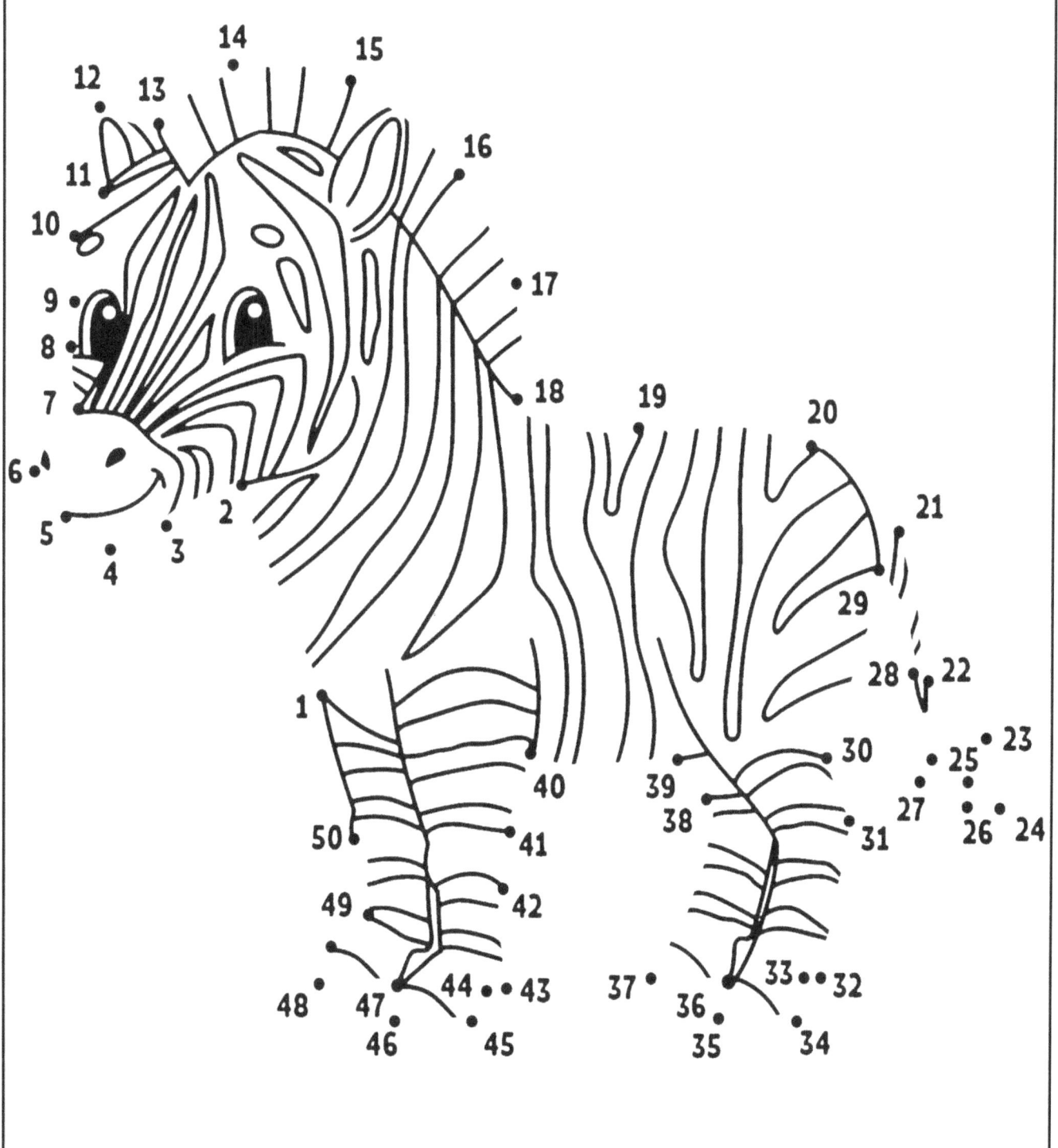

Test Your Color

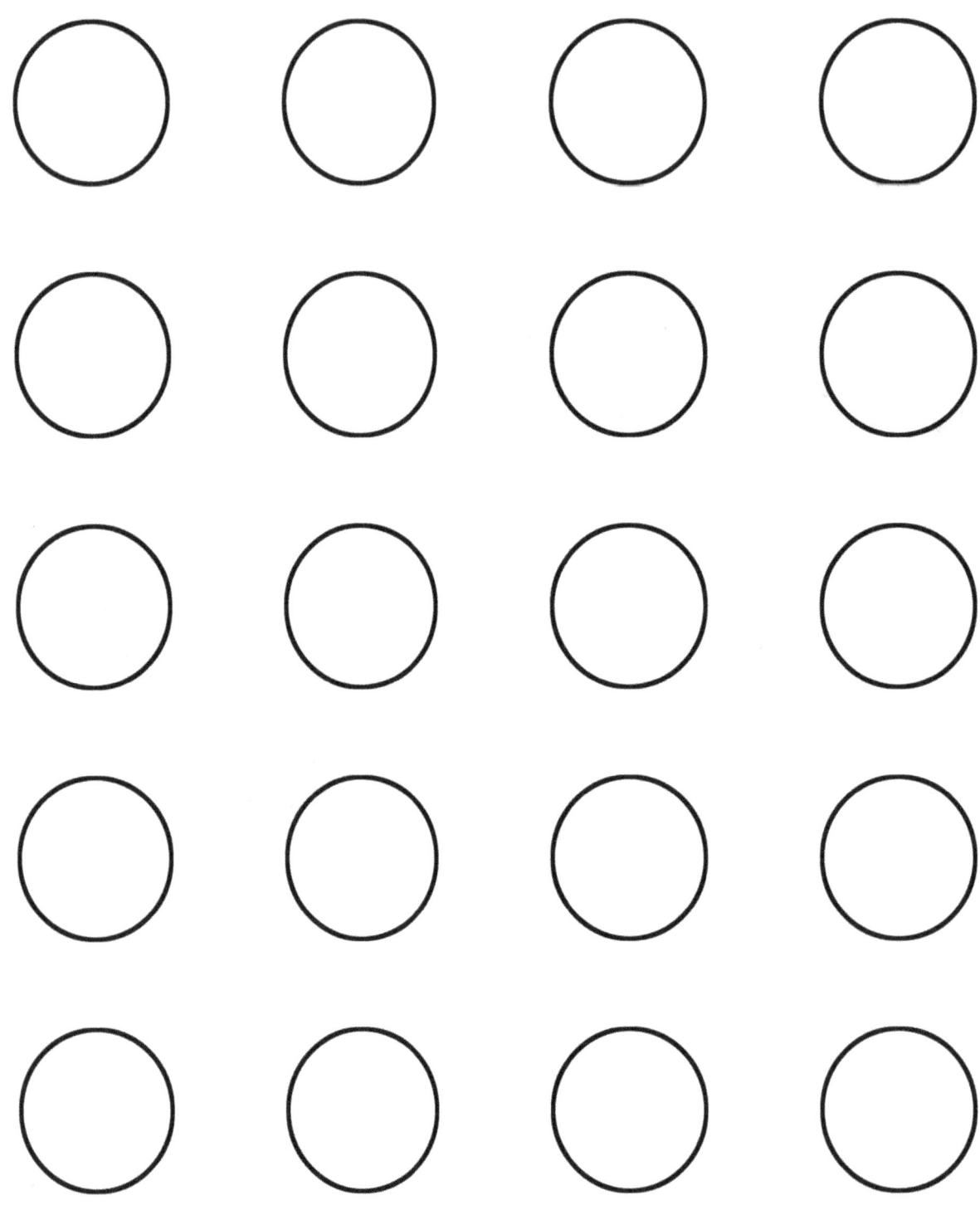

Connect the dots

Test Your Color

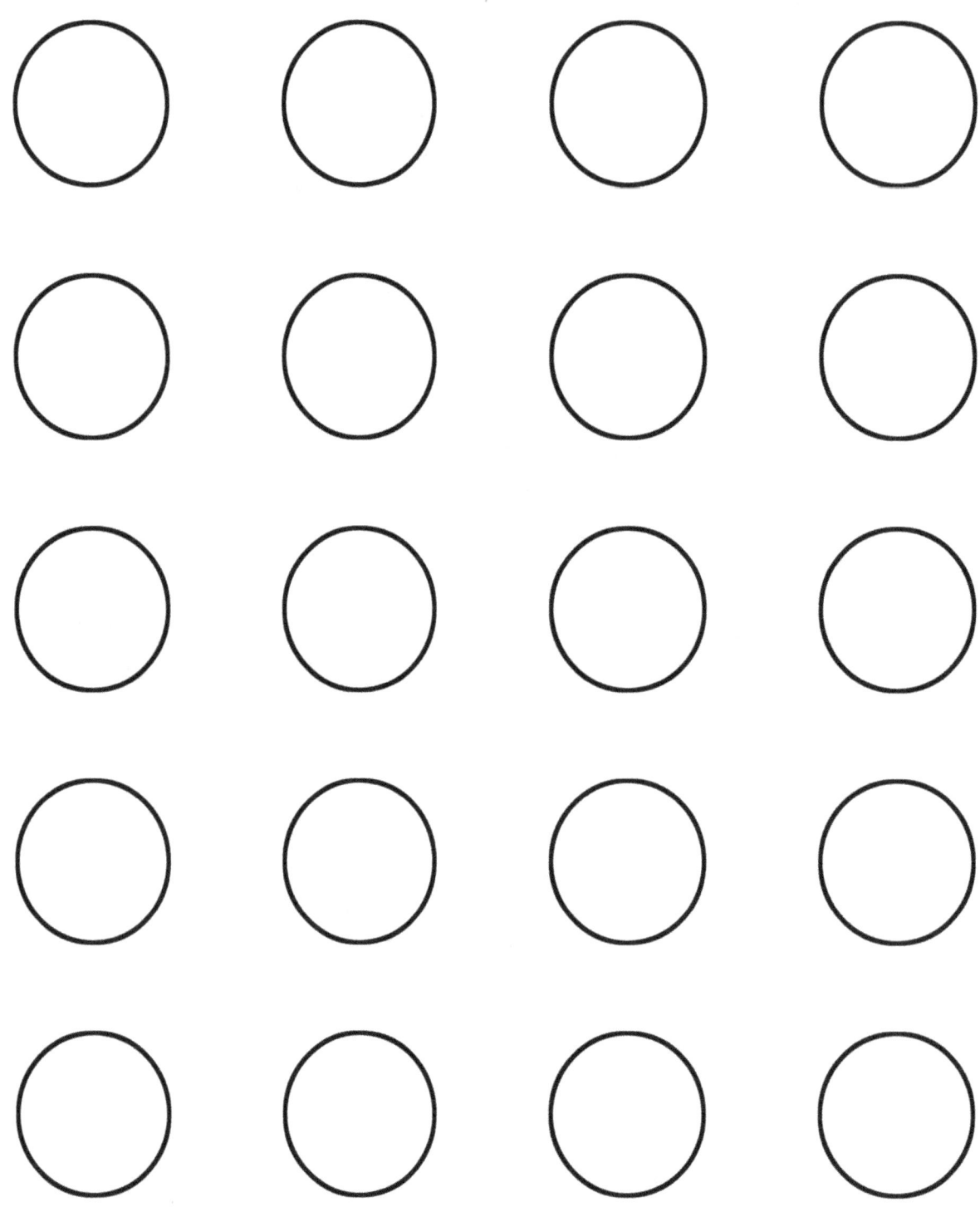

Connect the dots

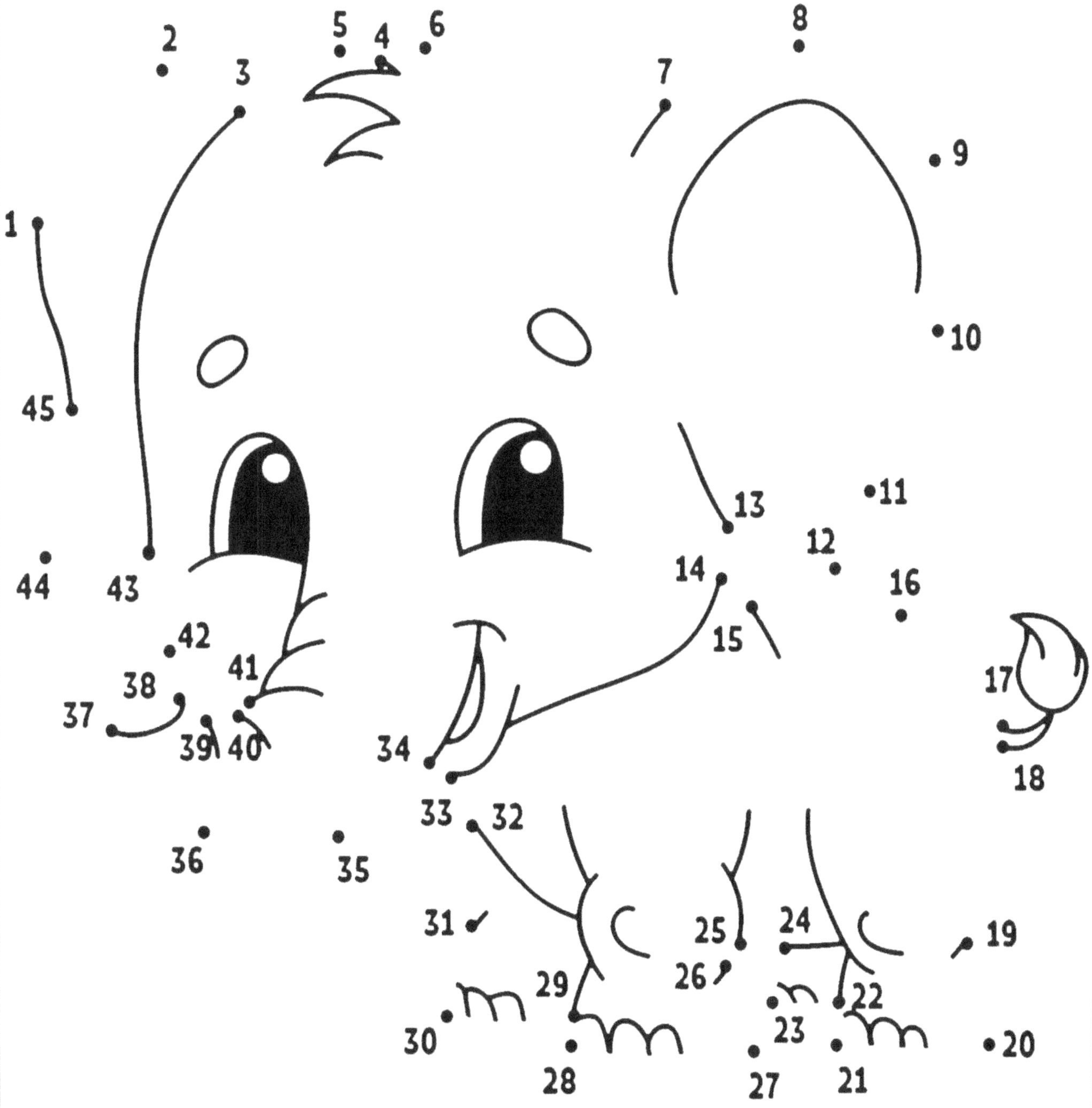

2

5 4 6

3

8

7

9

1

10

45

13

11

12

44

16

43

14

42

15

38 41

17

37 39 40

34

18

33 32

36 35

31

25 24

19

29

26

22

30

23

20

28

27 21

Test Your Color

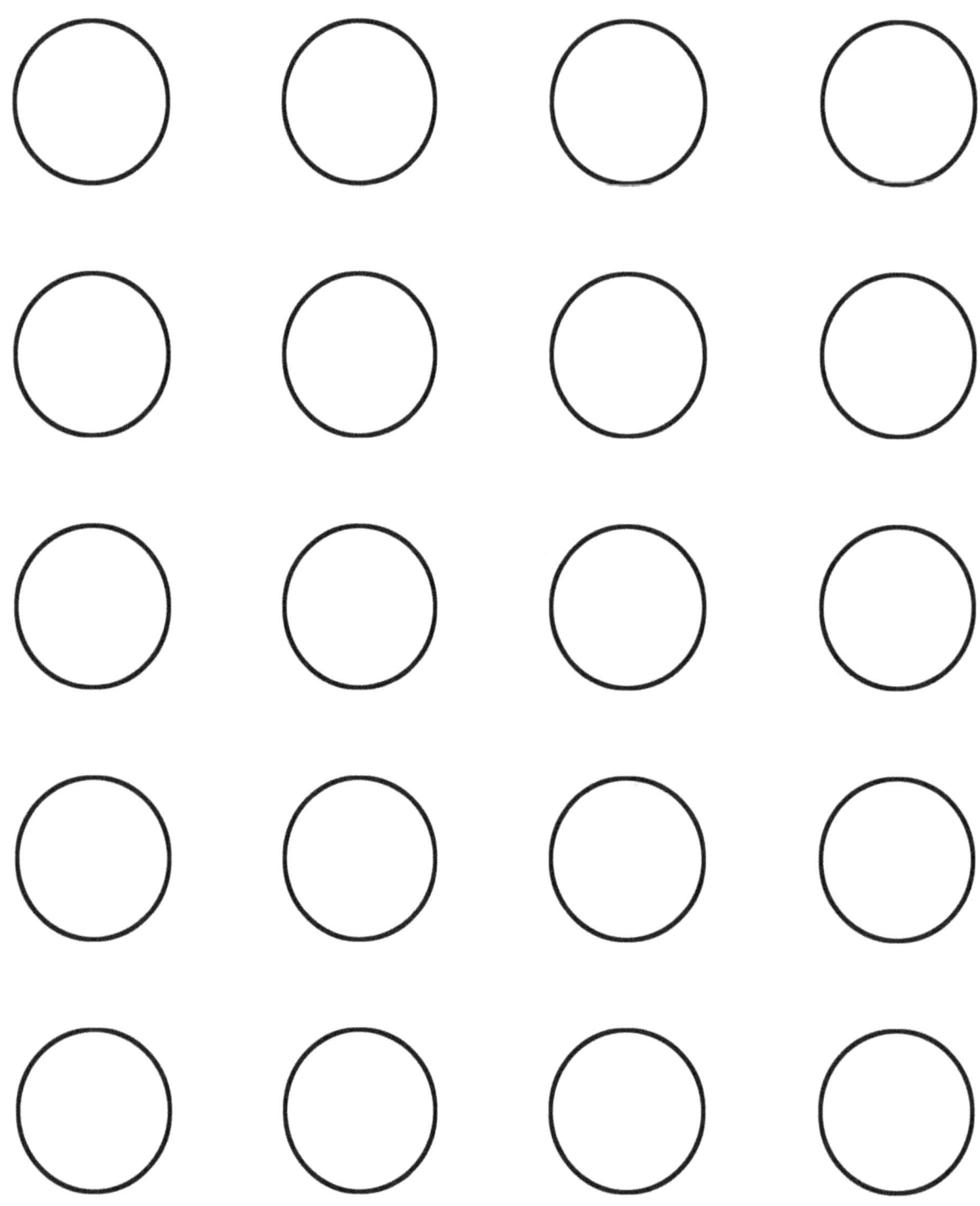

Test Your Color

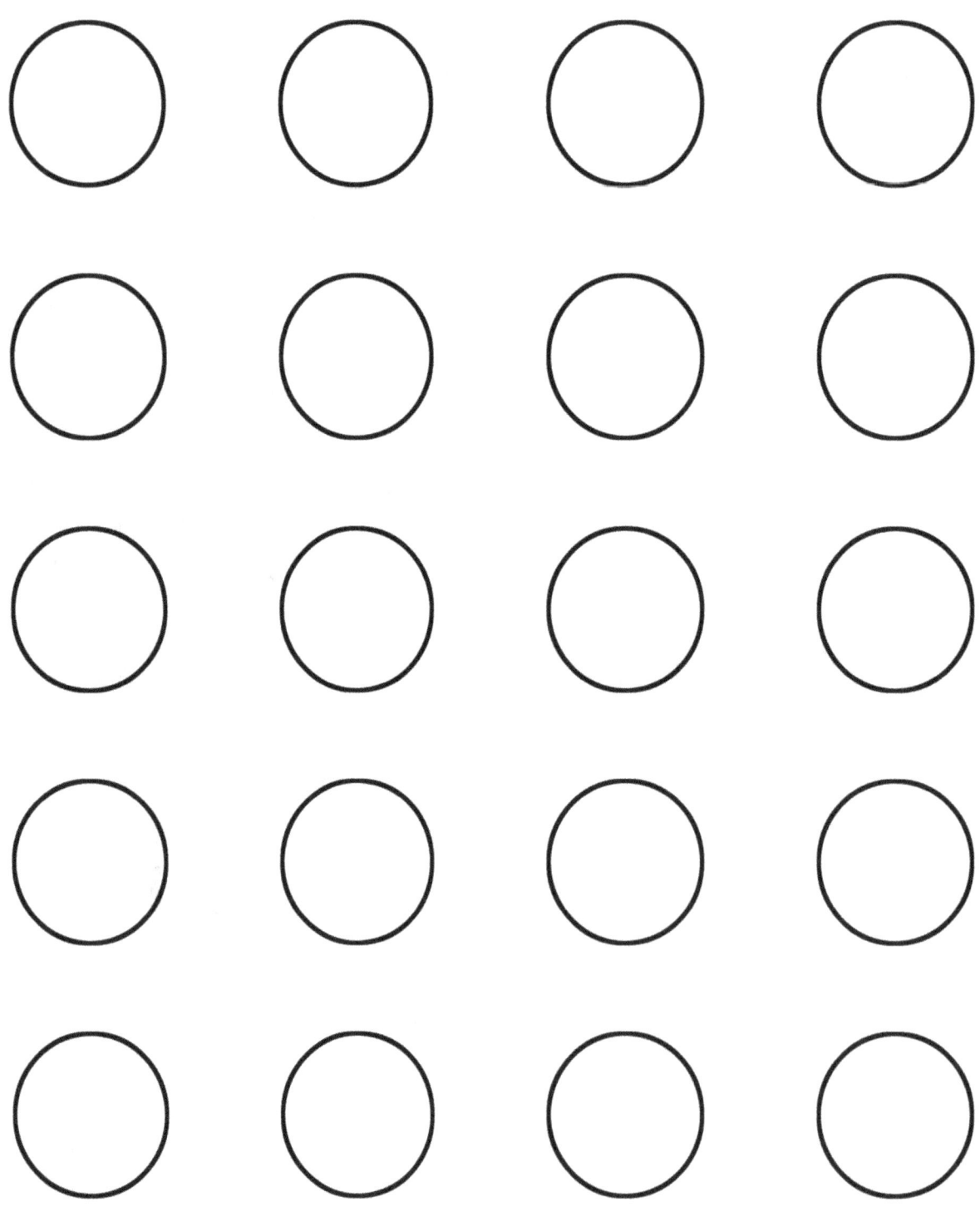

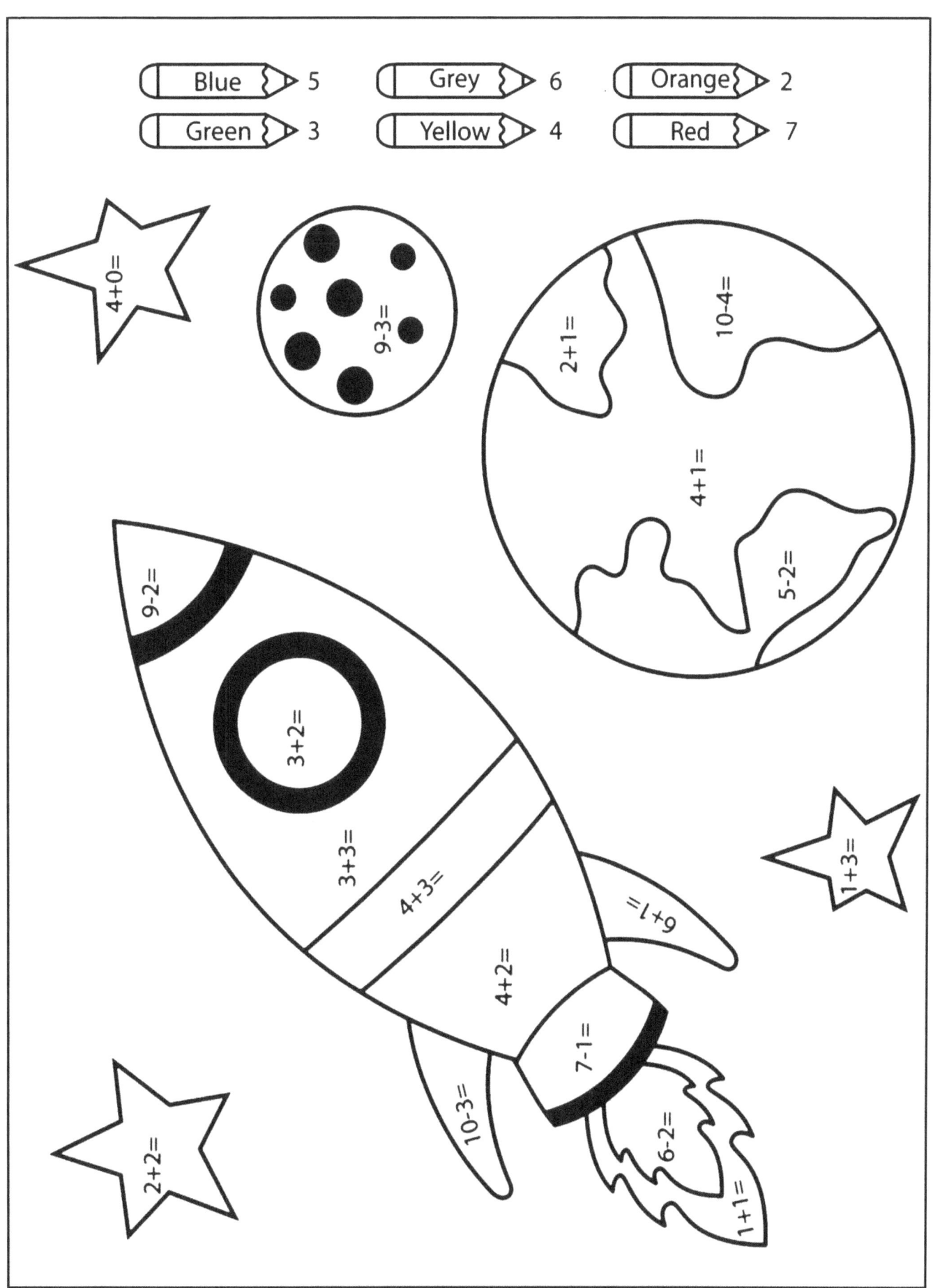

Test Your Color

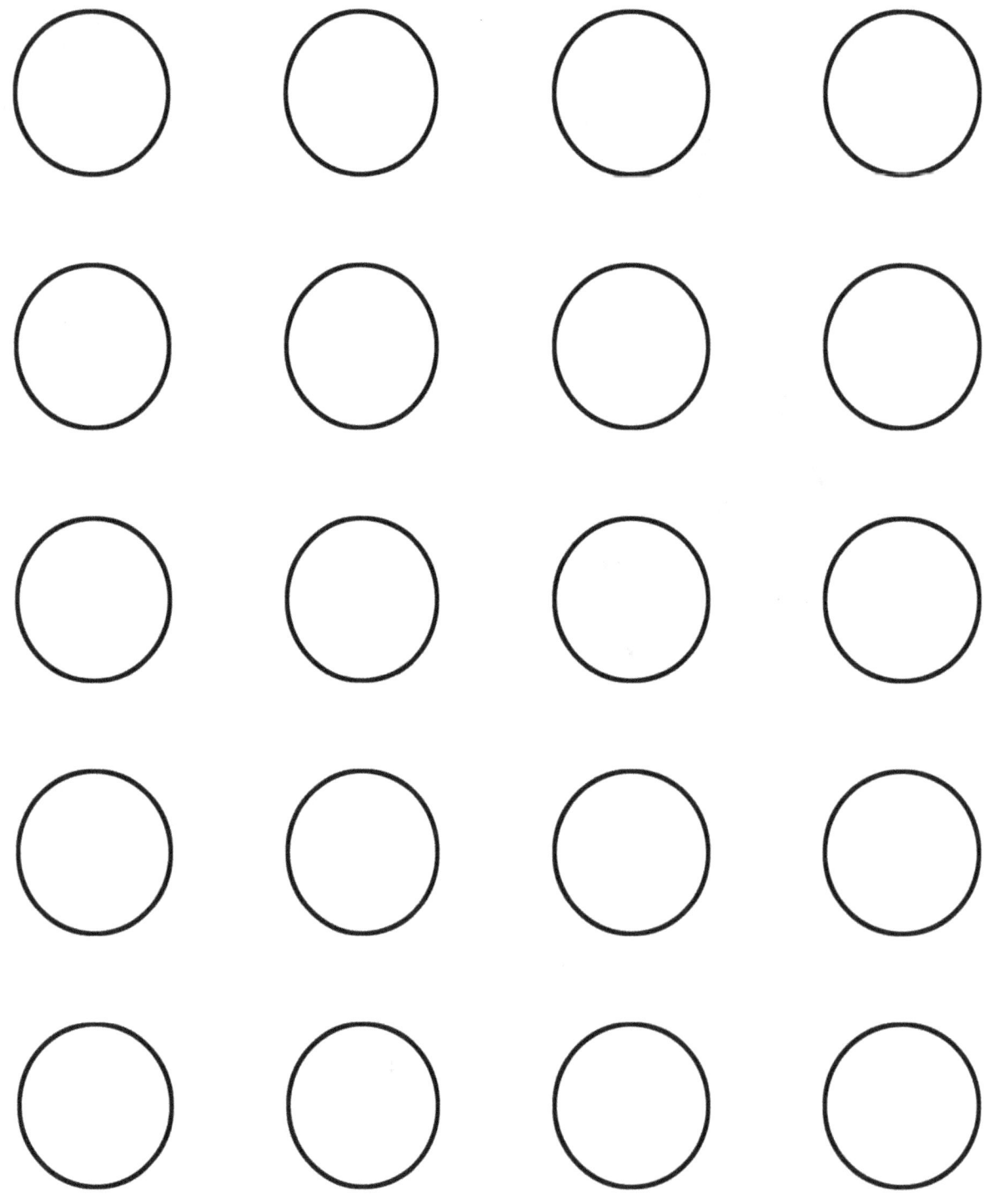

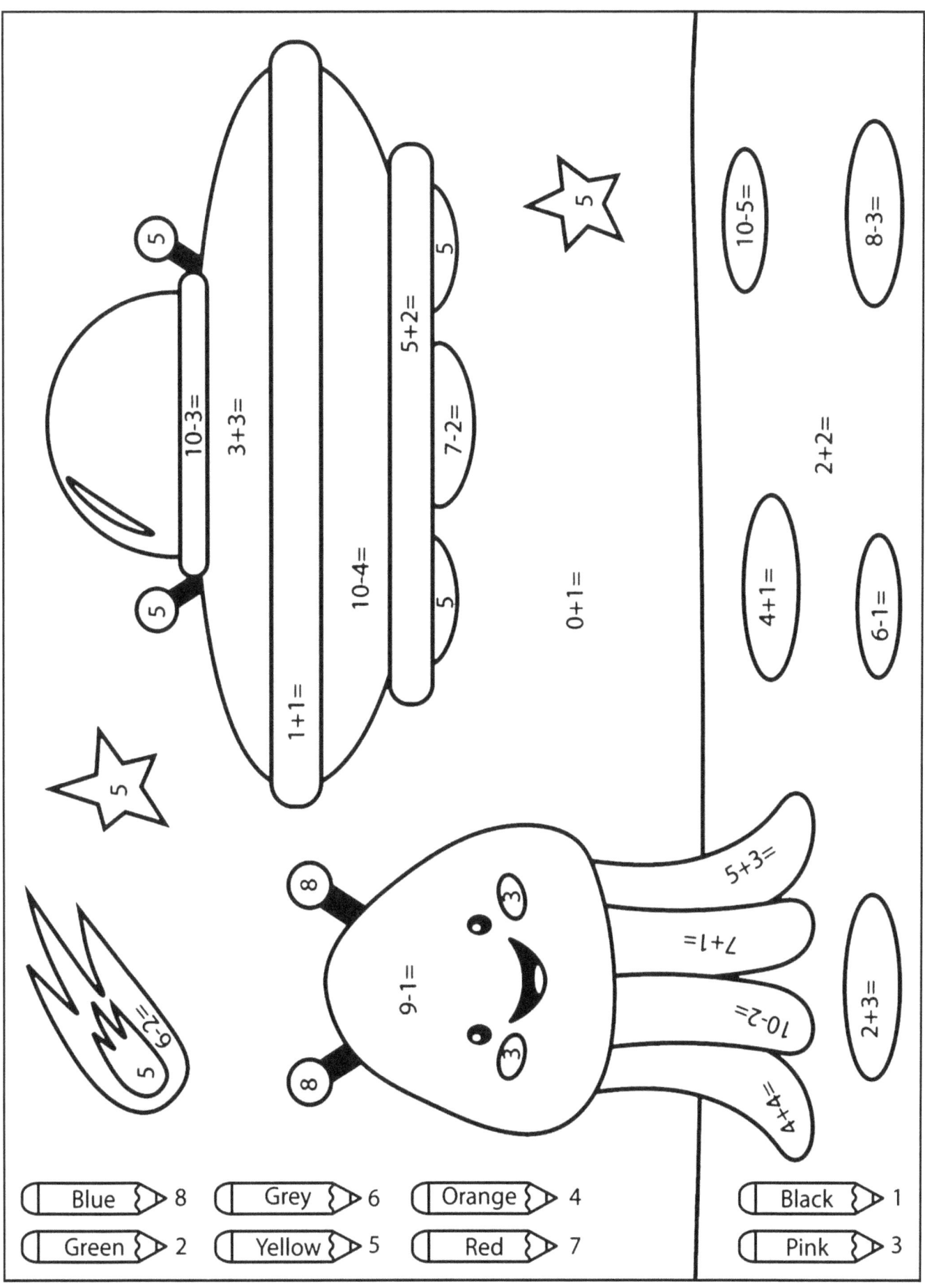

Blue ⊳ 8   Grey ⊳ 6   Orange ⊳ 4   Black ⊳ 1
Green ⊳ 2   Yellow ⊳ 5   Red ⊳ 7   Pink ⊳ 3

Test Your Color

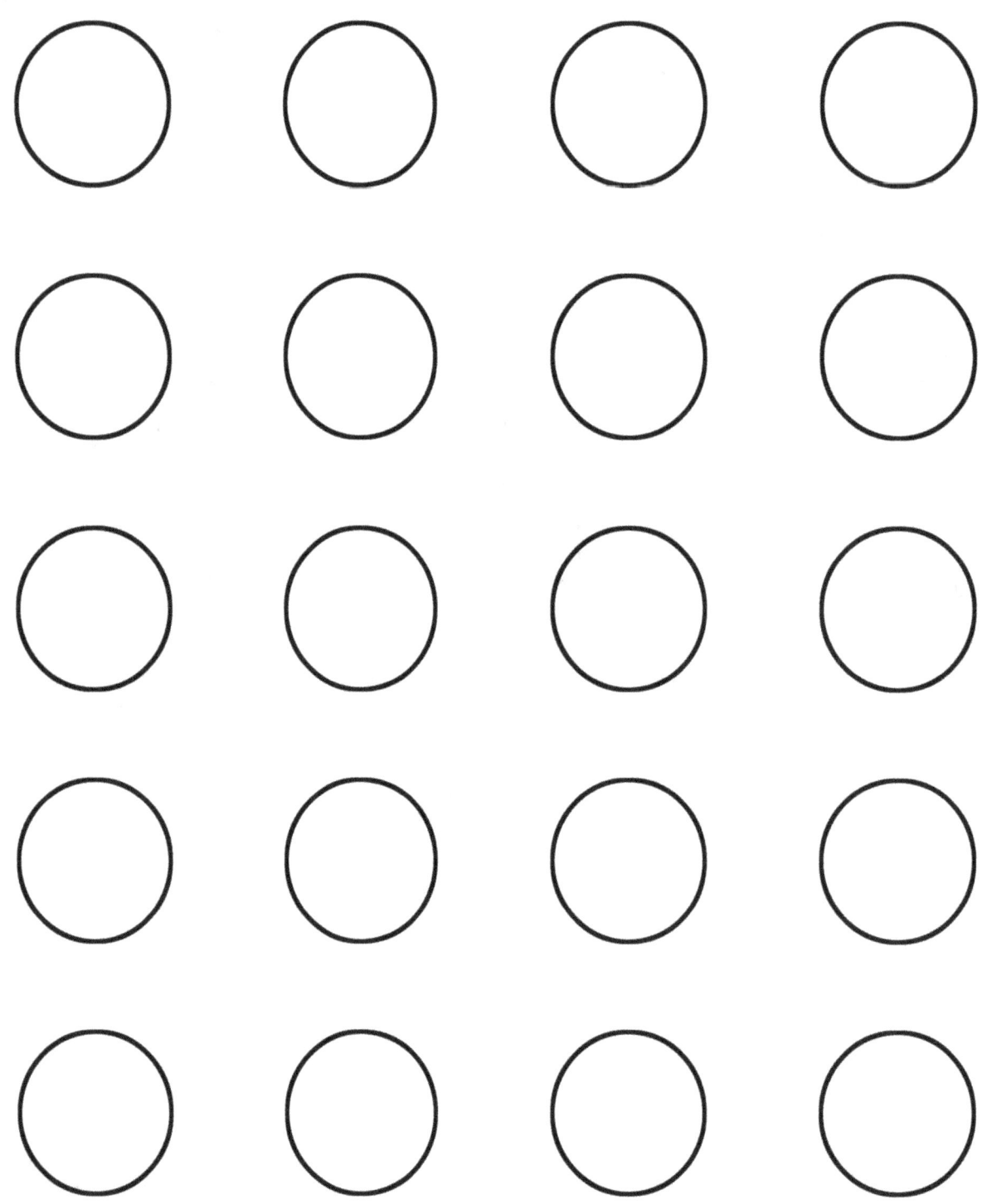

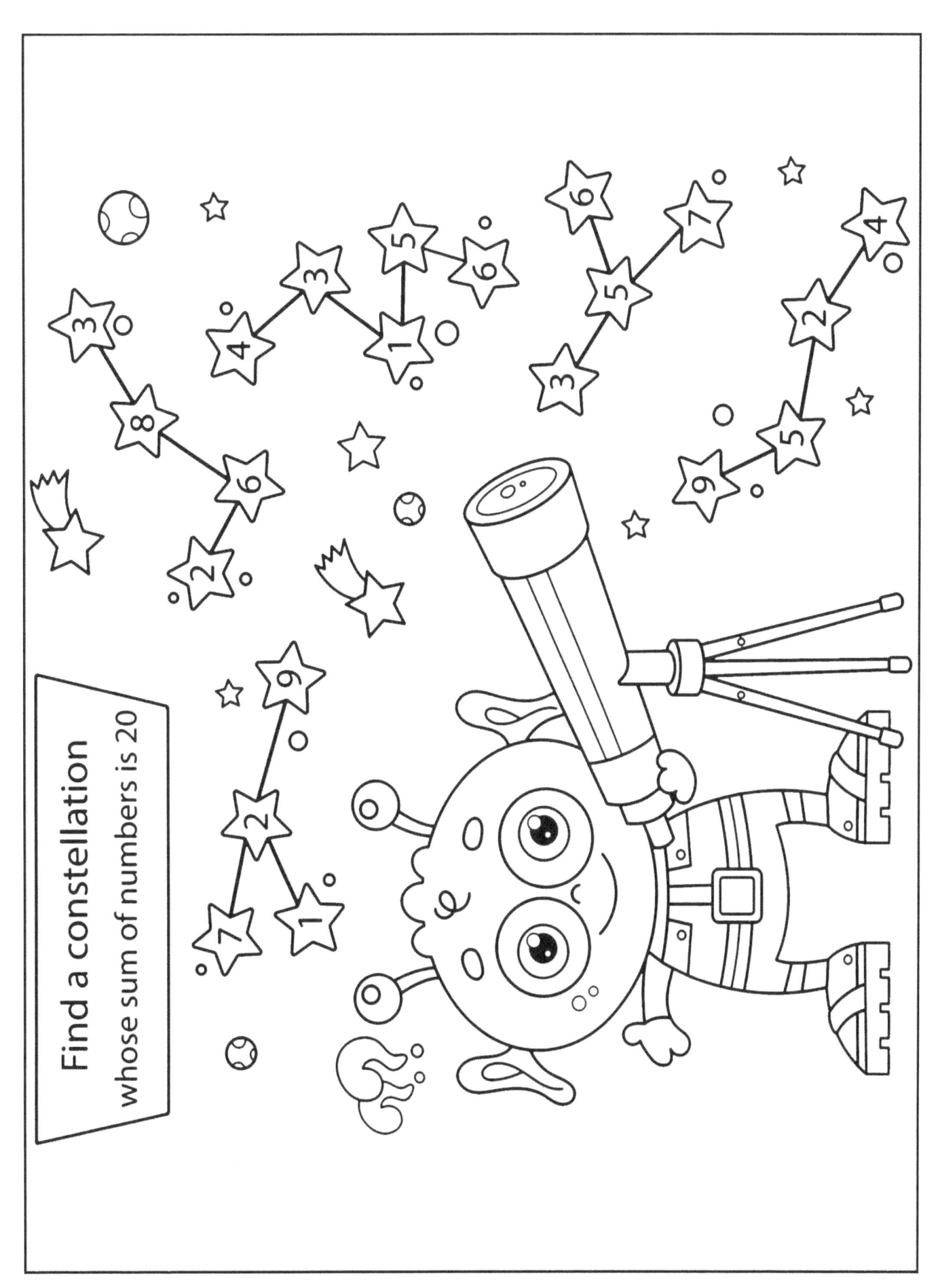

Find a constellation
whose sum of numbers is 20

Test Your Color

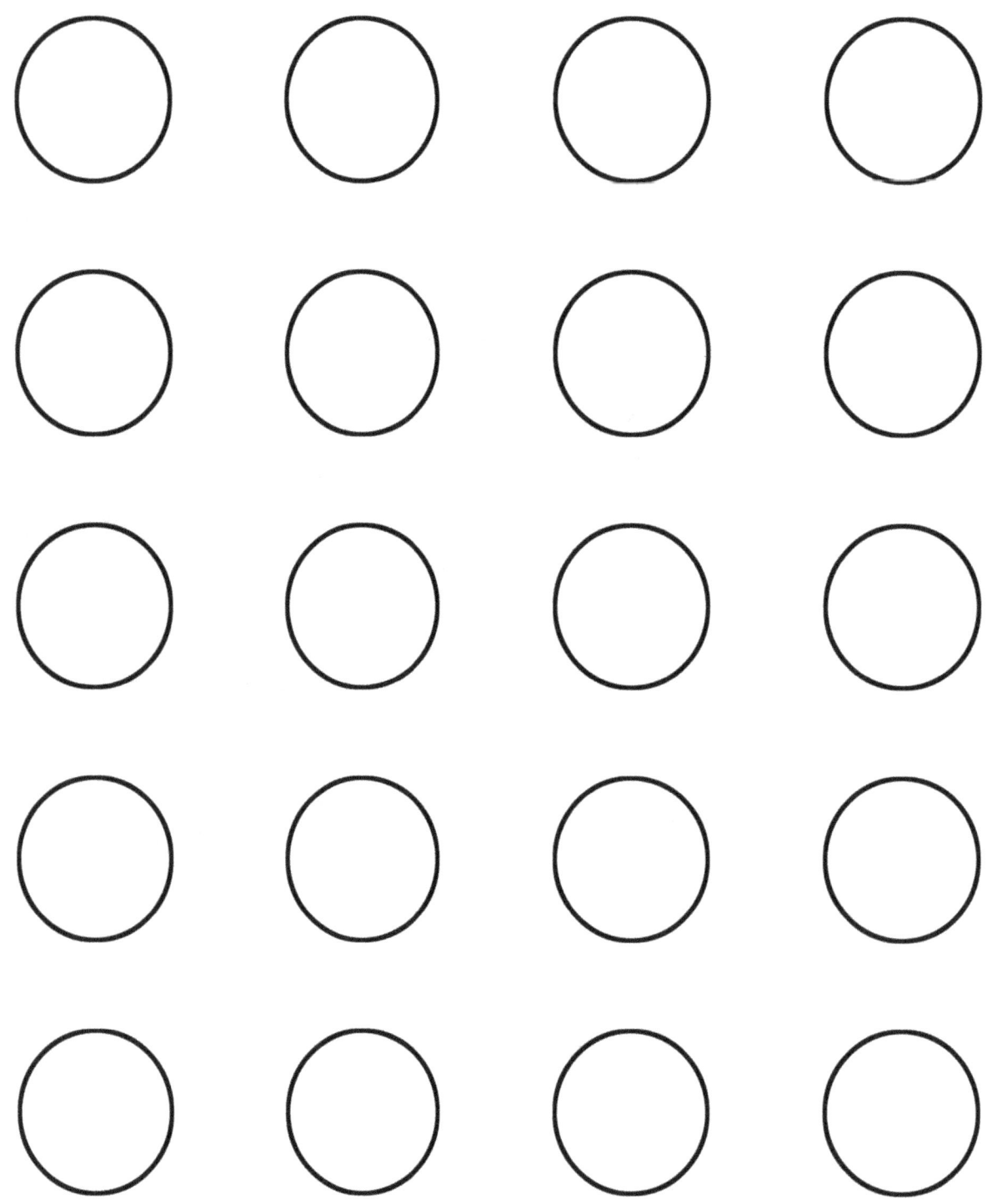

1

Test Your Color

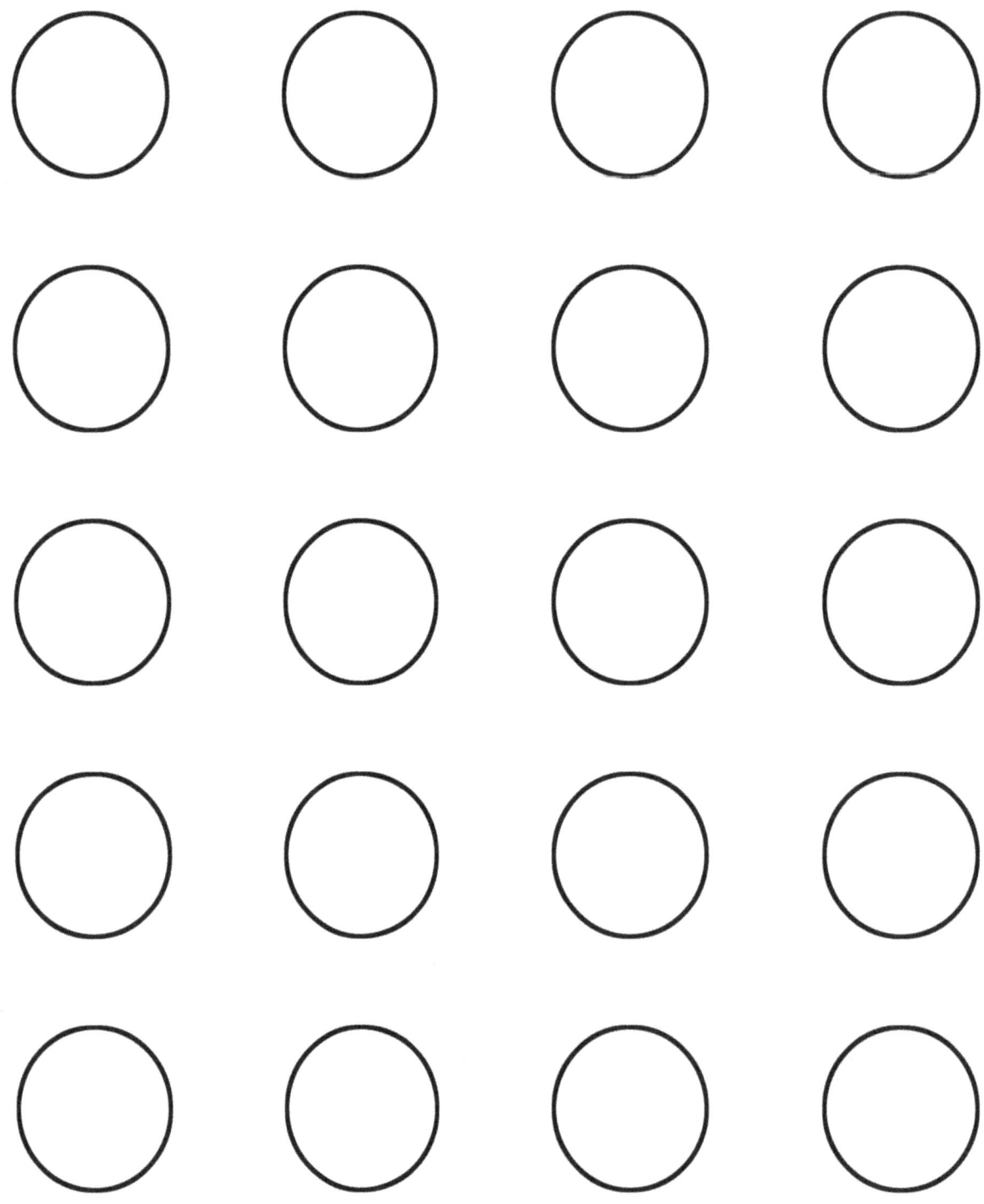

Test Your Color

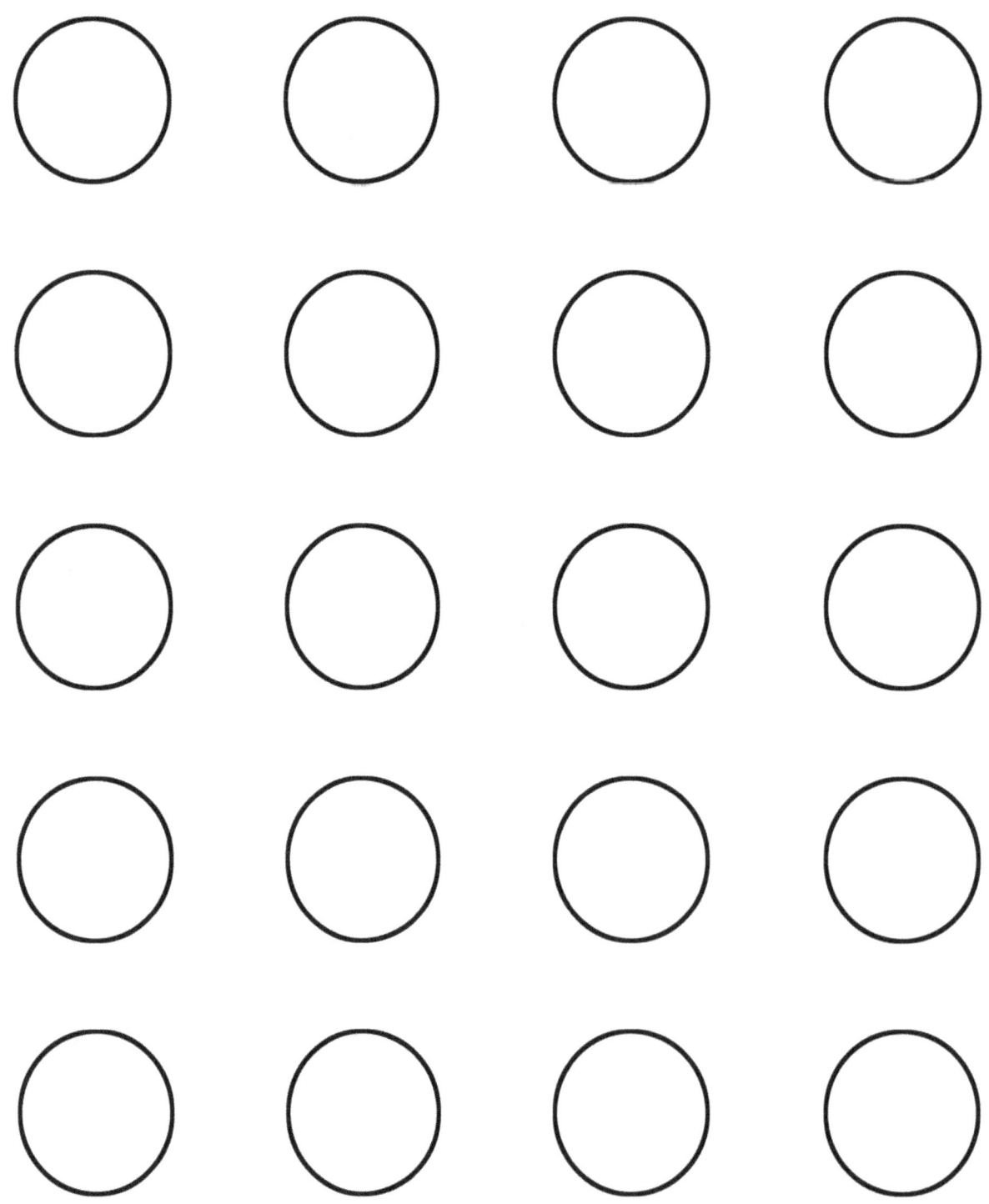

Test Your Color

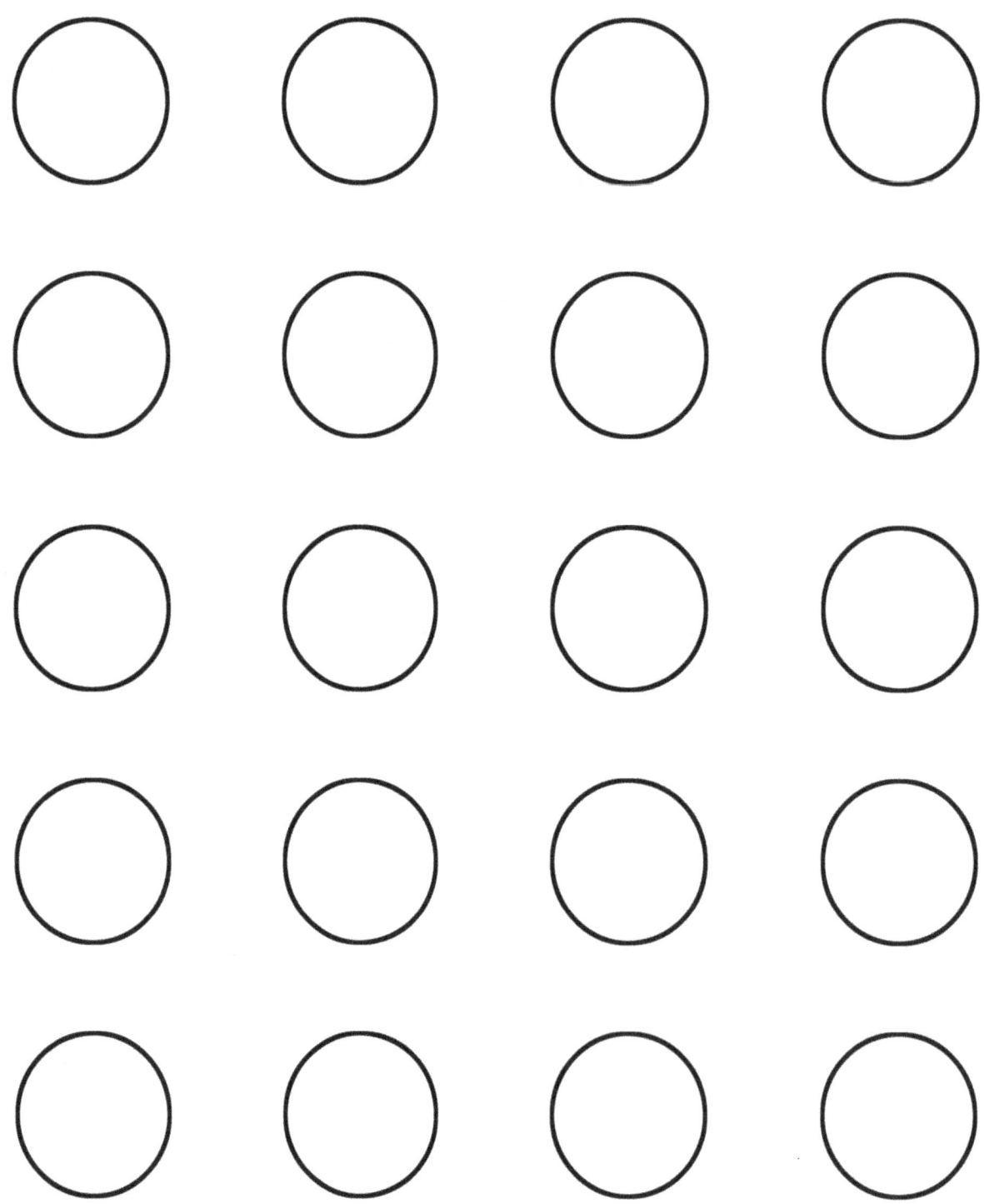

Test Your Color

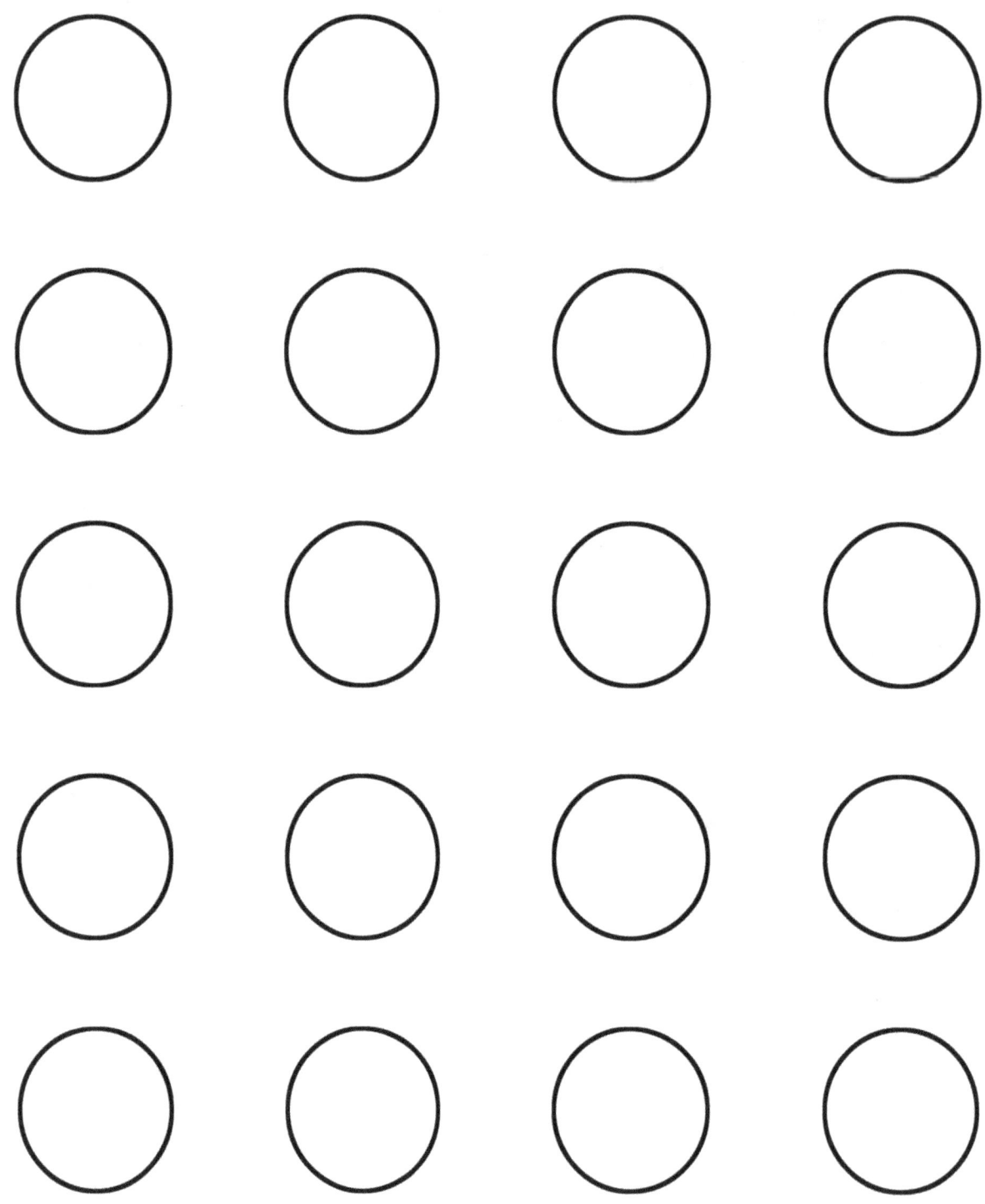

Test Your Color

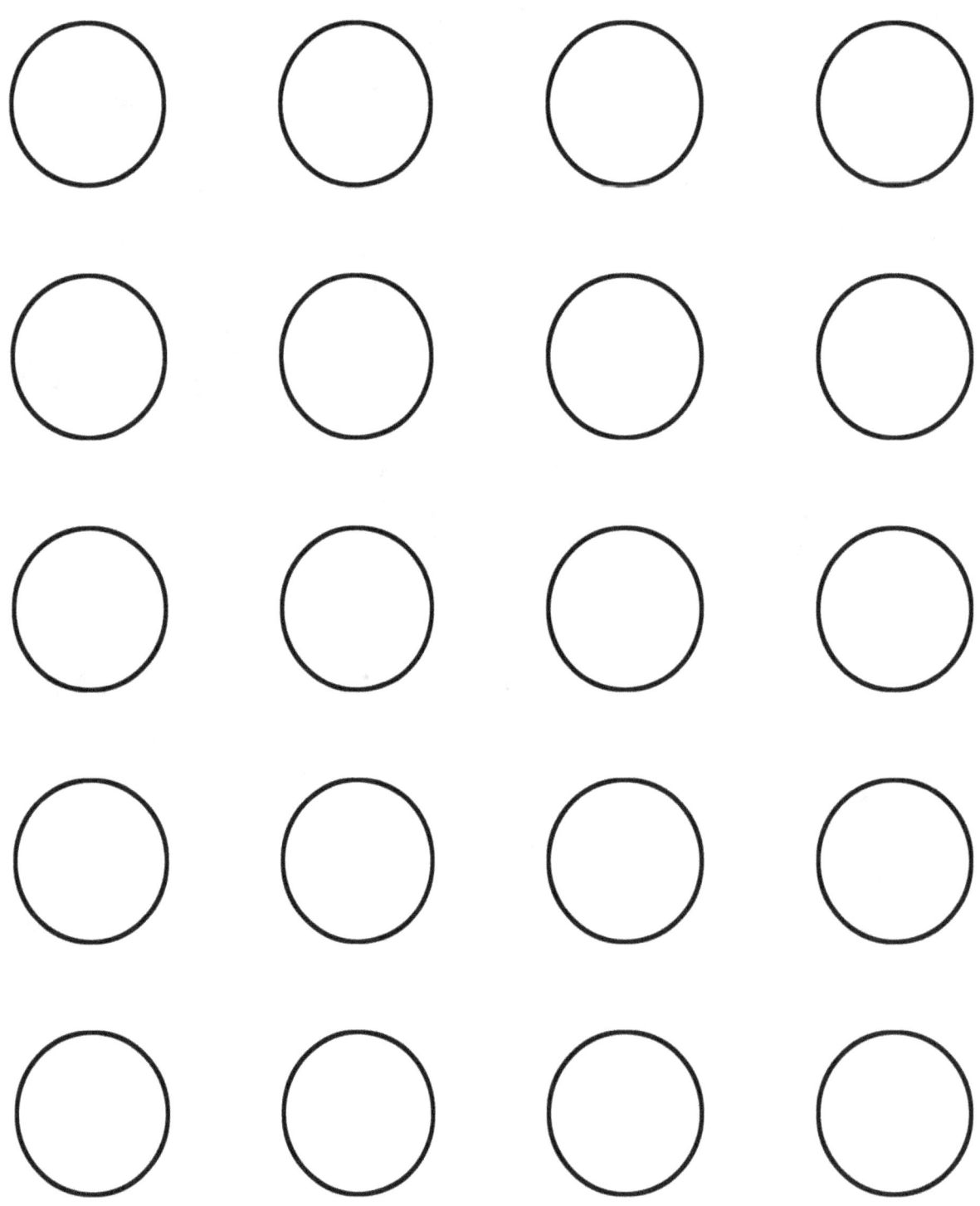

Test Your Color

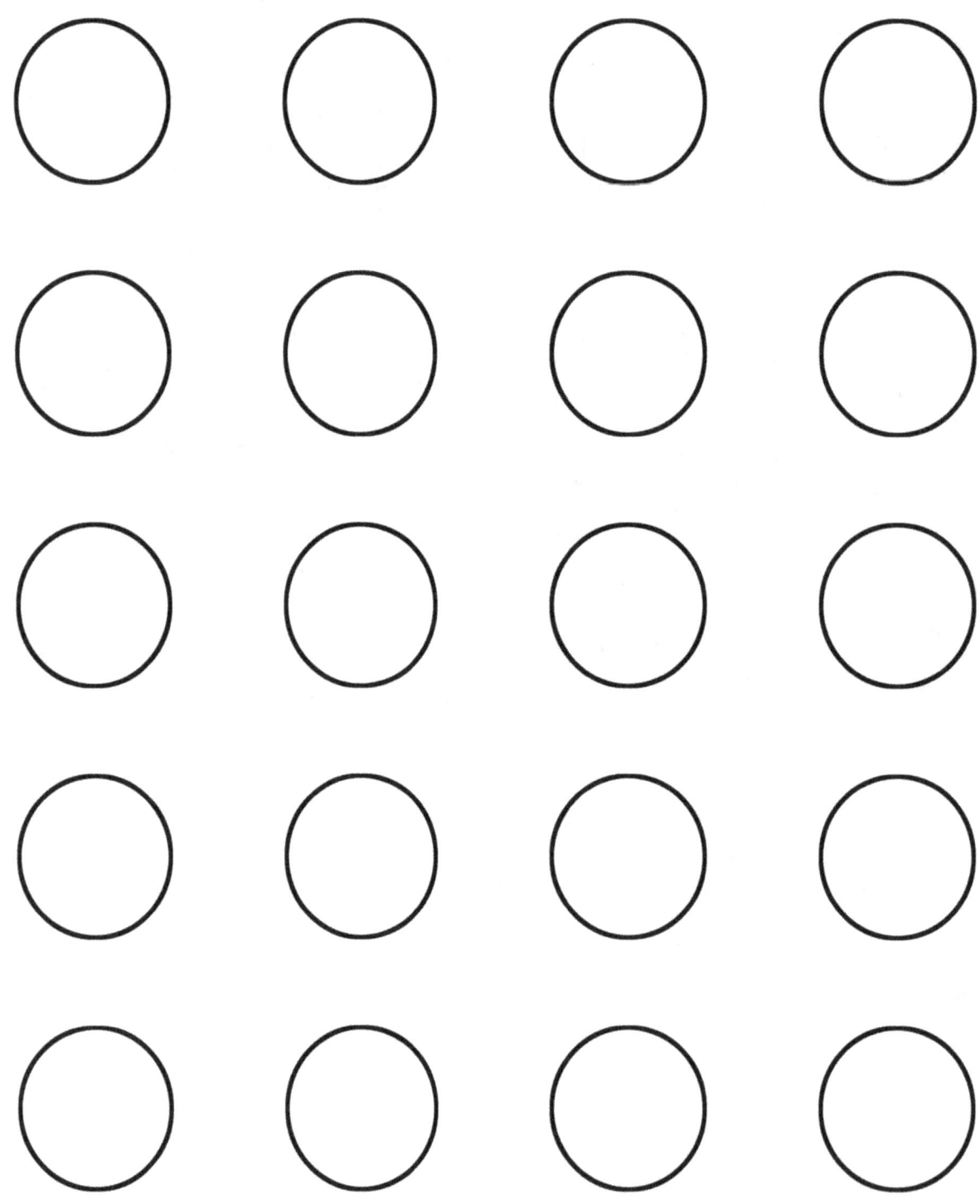

Test Your Color

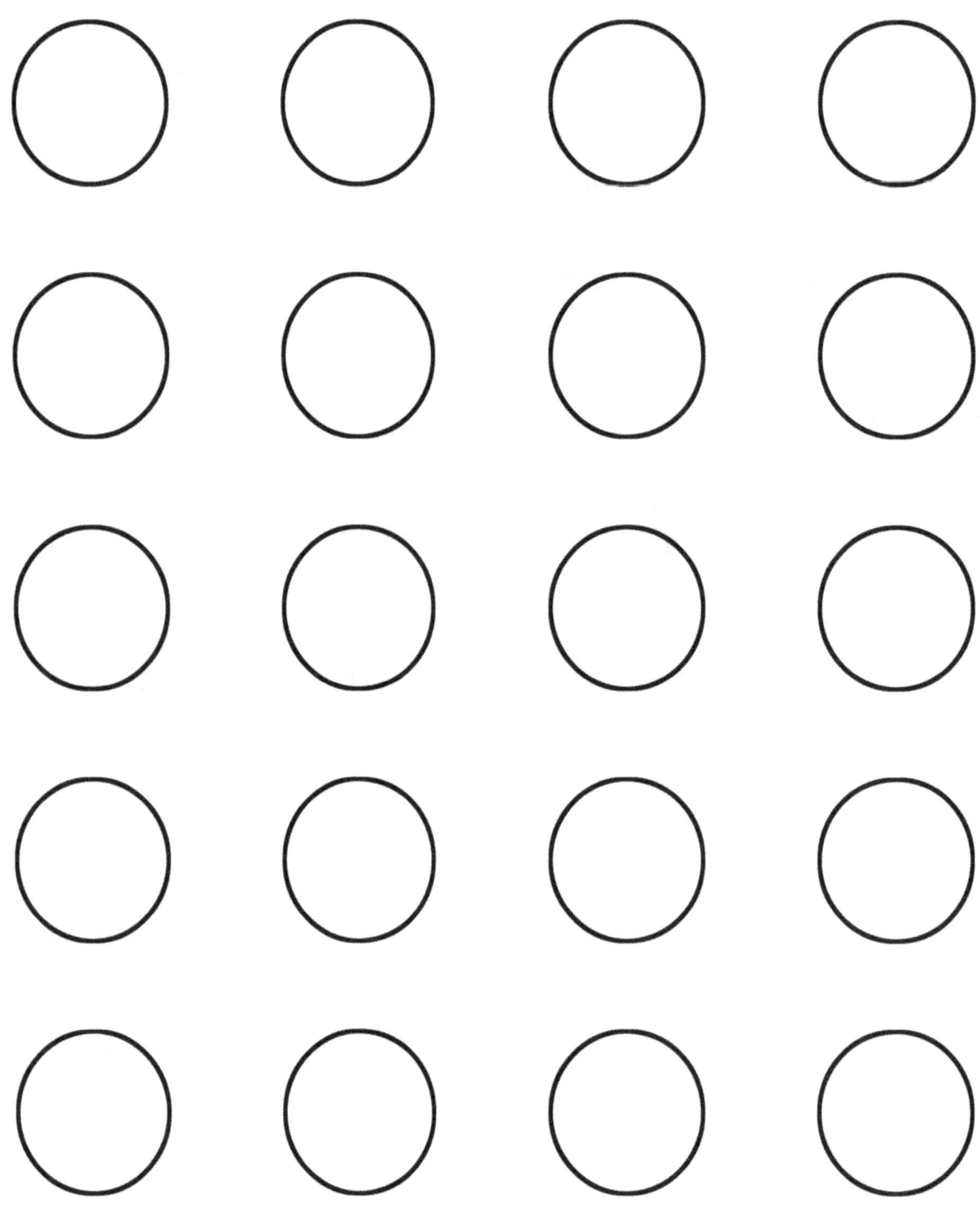

8

Test Your Color

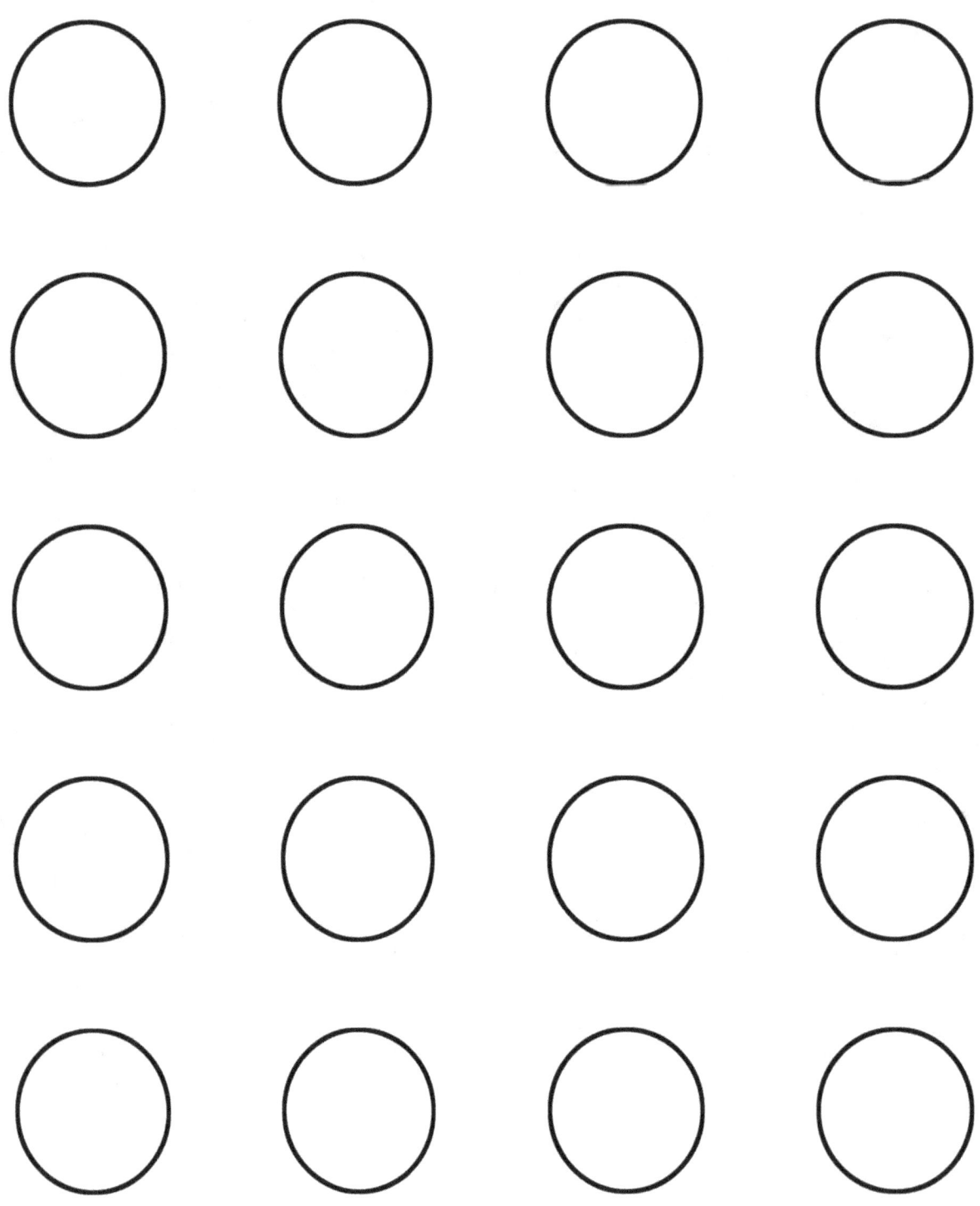

Test Your Color

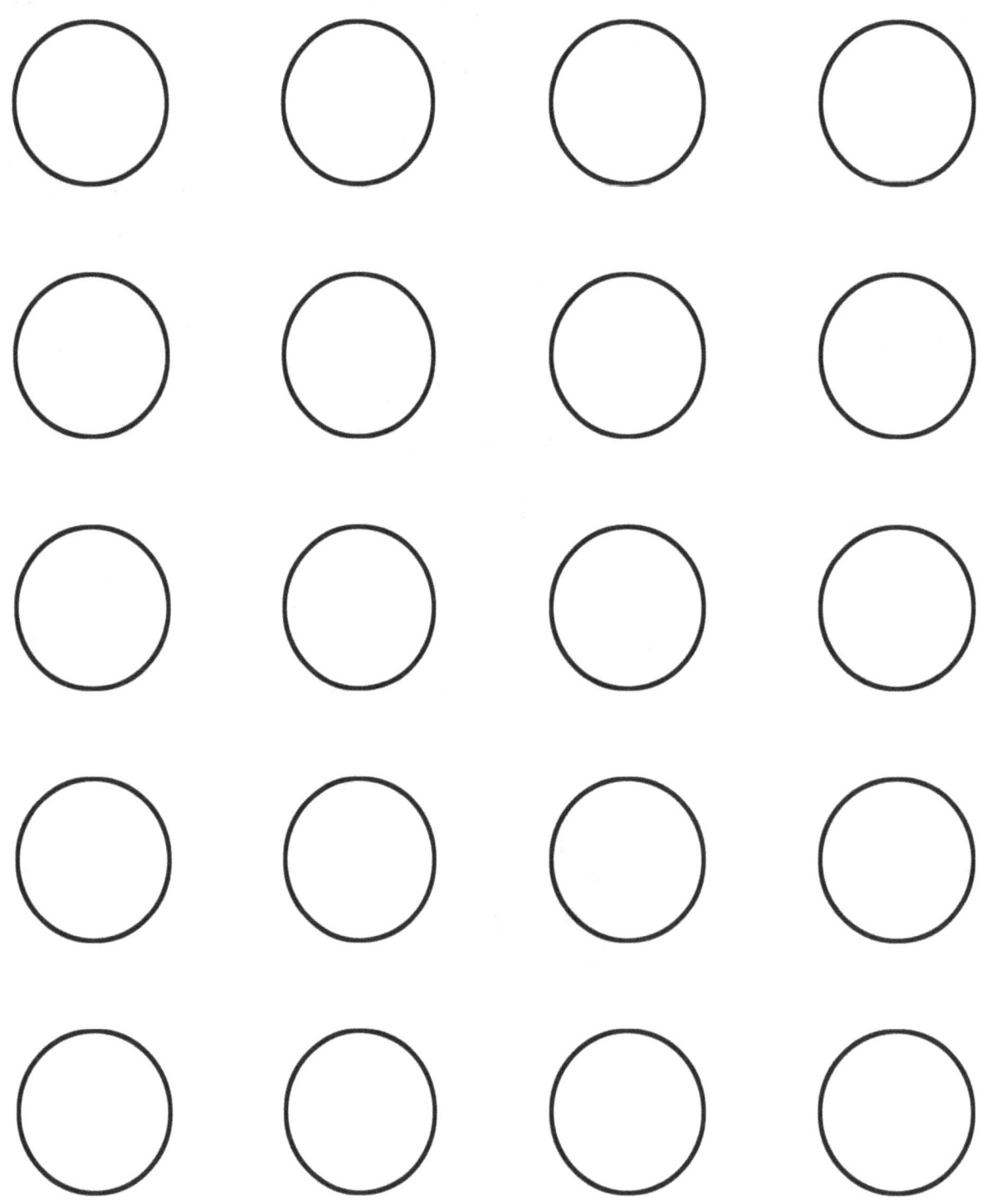

Test Your Color

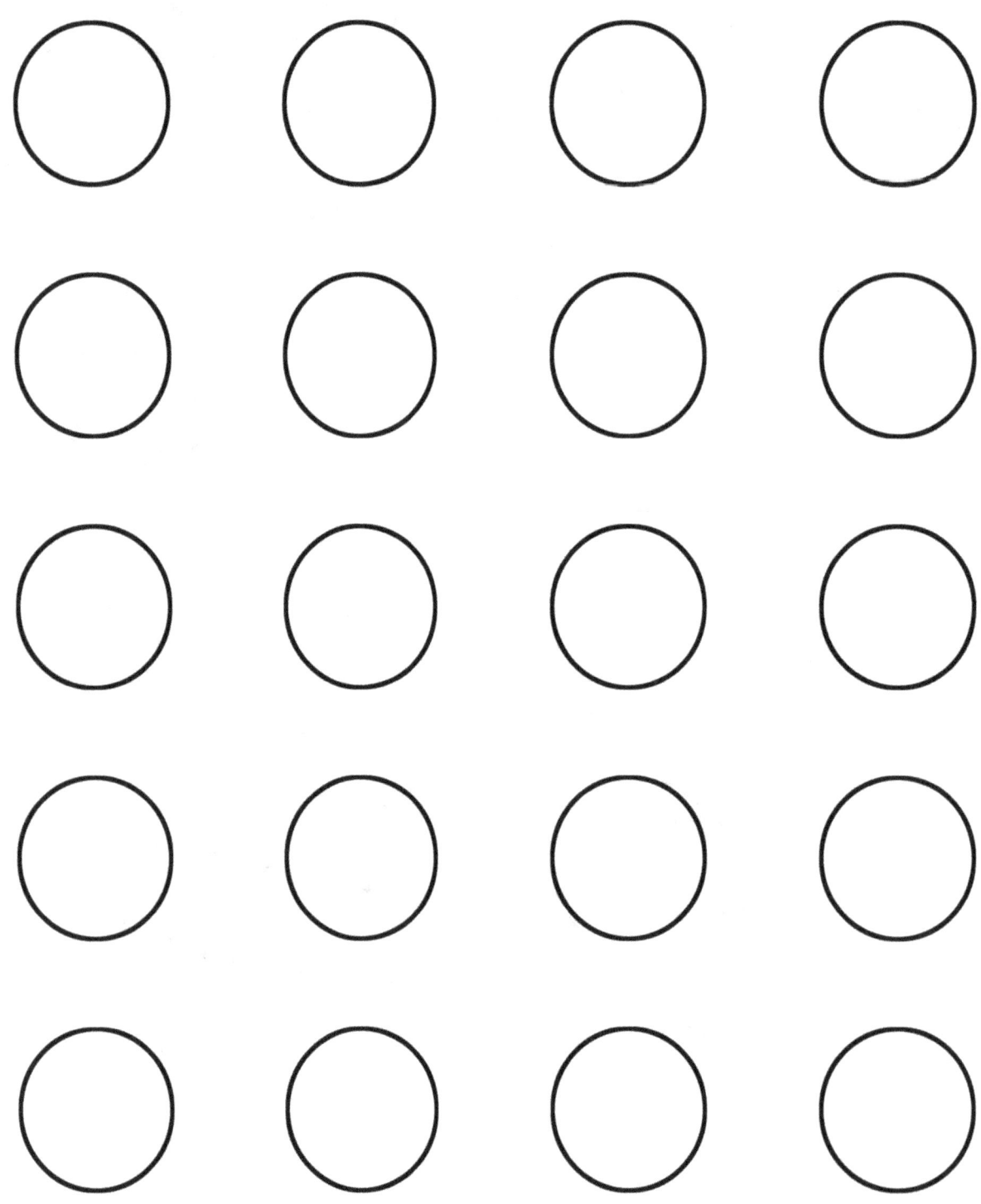

Test Your Color

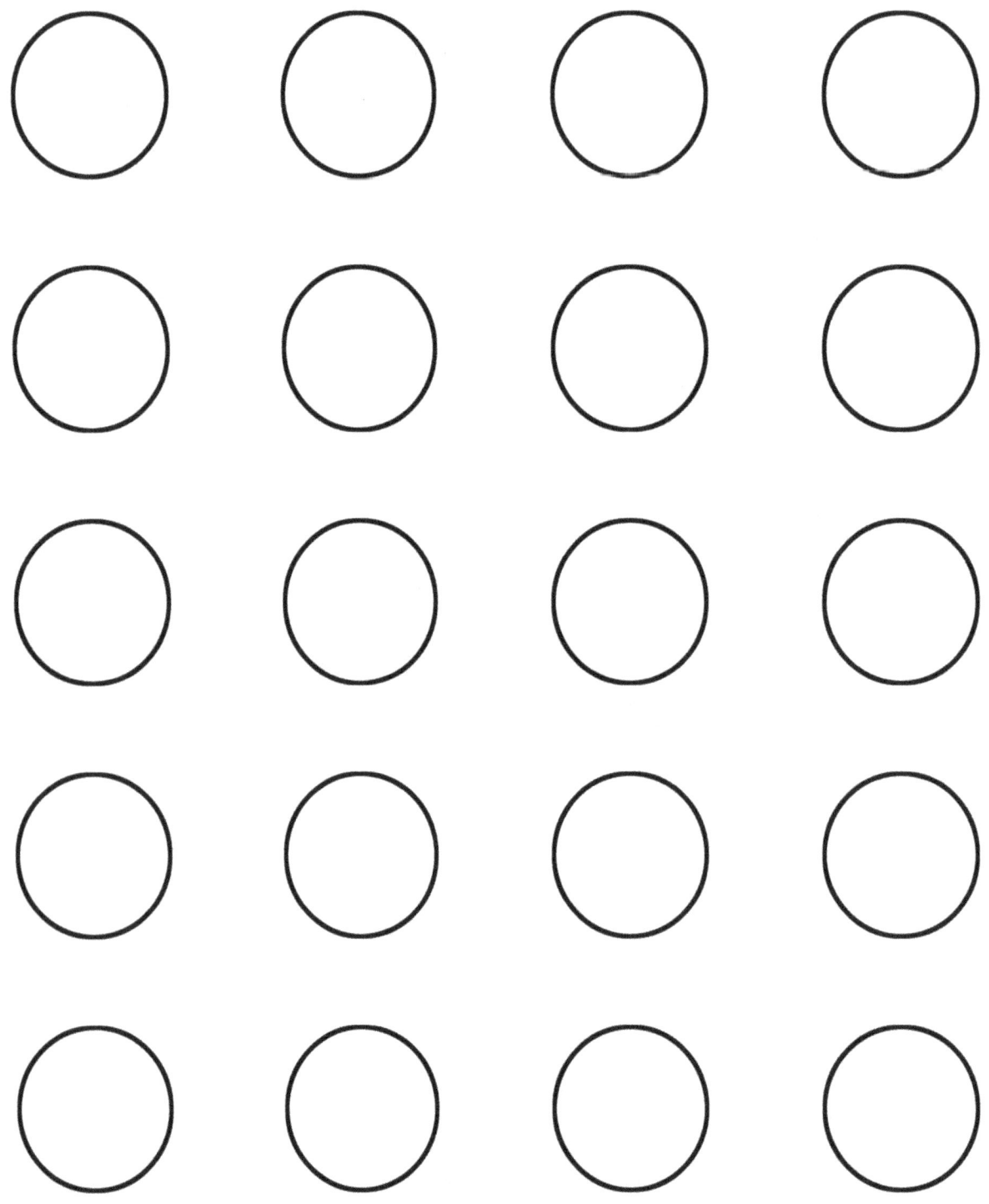

www.ingramcontent.com/pod-product-compliance
Lightning Source LLC
Chambersburg PA
CBHW080826220526
45467CB00008B/2201